GLASGOW UNIVERSITY PUBLICATIONS
XIX

THE
FABRICIAN TYPES OF INSECTS
IN THE HUNTERIAN COLLECTION AT
GLASGOW UNIVERSITY

THE
FABRICIAN TYPES OF INSECTS
IN THE HUNTERIAN COLLECTION AT
GLASGOW UNIVERSITY

COLEOPTERA
PART I

By

ROBERT A. STAIG, M.A., F.R.S.E.

LECTURER IN ZOOLOGY (ENTOMOLOGY)
UNIVERSITY OF GLASGOW

CAMBRIDGE
AT THE UNIVERSITY PRESS
1931

CAMBRIDGE UNIVERSITY PRESS
Cambridge, New York, Melbourne, Madrid, Cape Town,
Singapore, São Paulo, Delhi, Mexico City

Cambridge University Press
The Edinburgh Building, Cambridge CB2 8RU, UK

Published in the United States of America by Cambridge University Press, New York

www.cambridge.org
Information on this title: www.cambridge.org/9781107685475

First published 1931
First paperback edition 2013

A catalogue record for this publication is available from the British Library

ISBN 978-1-107-68547-5 Paperback

CONTENTS

ILLUSTRATIONS

INTRODUCTION

The *Hunterian Collection of Insects* represents a small but very important part of the extensive Museum formed by *Doctor William Hunter* who was Court Physician to Queen Charlotte, consort of George III.

It is well known that Dr Hunter intended to found, during his lifetime, a School of Anatomy and Medicine in London; he was prepared to establish it at his own expense and to endow it with his Museum and Library, provided the Government would grant a suitable site on Crown land. His generous offer was not entertained.

Hunter then thought of Glasgow, his Alma Mater; writing about his plans to his old friend, Dr Cullen, he said, "I have a great inclination to do something considerable at Glasgow some time or other". Again, however, he suffered disappointment; for at that time (1765), as it happened, Cullen and others, who were essential to the success of his cherished scheme, had left or were about to leave Glasgow.

Ultimately, Dr Hunter bequeathed his valuable collections to Glasgow University, all his anatomical preparations, his zoological, geological and ethnological specimens, his unique library of books and manuscripts, his magnificent collection of coins and various art treasures. In his will he assigned the usufruct of these collections for a period of thirty years to his nephew and heir, Dr Matthew Baillie, brother of Joanna Baillie. The collections were received by Glasgow University in 1807.

Hunter was not, like most collectors of his day, and some private collectors of the present time, a mere gatherer of curiosities or natural rarities...every specimen or preparation suggested to him either a fact recorded or a problem to be solved. The last thing he would have approved was a *dead* museum, a mere storehouse. He surely and certainly meant that his preparations and his specimens were to take their place with others as stones built into the rising house of knowledge[1].

With that aim ever in view, the Hunterian Collections have been greatly enriched, more especially in recent years, by many valuable acquisitions. Dr Hunter's Cabinets of Insects have formed a great nucleus around which there is now a considerable amount of material for reference and instruction in Entomology, the latest and most notable additions being the extensive collections of British and Exotic Coleoptera which belonged to the late Mr Thomas G. Bishop, and the great collection of British Insects of all Orders, the life work of Mr James J. F. X. King, F.E.S.

Some years ago (in 1910) Professor Graham Kerr published a paper[2] in which he made special reference to Dr Hunter's Insects. The particular interest and importance of this collection is mainly due to the fact that many of the specimens are the types of insect species founded by *Johann Christian Fabricius*, the great pioneer of Systematic Entomology in the eighteenth century.

Fabricius was (latterly) Professor of Natural History and Rural Economy in the University of Kiel; he was born in Tondern, Schleswig, on January 7th, 1745 and he died May 3rd, 1810. As a small boy he "collected plants and insects and studied the *Species Plantarum* and the *Philosophia Botanica* far more diligently than *Cornelius Nepos* or *Cicero*"; and when he was a little older, he developed an ardent desire to go to Upsala and become a pupil of the celebrated Linnaeus. Wisely recognising this early evidence of a future vocation, his father granted his wishes and sent him to Upsala at the age of seventeen. Daily intercourse with the great master

[1] *William Hunter and his Museum.* An Oration delivered in the University of Glasgow on Commemoration Day, 22nd June, 1922, by Professor T. H. Bryce, M.A., M.D., F.R.S. (Glasgow, MacLehose, Jackson & Co.).

[2] "Remarks upon the Zoological Collection of the University of Glasgow" by Professor J. Graham Kerr (*Glasgow Naturalist*, Vol. II, No. 4, September, 1910).

soon determined his bent of mind, and Fabricius then began to devote himself to the study of Insects, thenceforth his main interest in life.

After two years with Linnaeus, he returned to Copenhagen and set about compiling the *Genera Insectorum*, in accordance with the small collection he then possessed; this was the foundation of his first notable work, the *Systema Entomologiae*. About that time he began those journeys, in many parts of Europe, which throughout his life he voluntarily undertook for the advancement of Entomology. He visited the various towns, became acquainted with the leading men of science, more especially those who studied and collected insects, and thus gained access to most of the noted collections in Europe.

In his *Autobiography*[1] Fabricius mentions his indebtedness to Dr Solander, of the British Museum, who introduced him to Hunter, Banks, Drury and many others in London "whose houses and libraries and collections were soon opened to me. I determined and described the insects, and arranged the species of the collections. My *System of Insects* gained ground considerably, as well by the more exact definition of the species, as by the addition of a considerable number of genera".

The following passage, in which he refers to a later London visit (1772–1775), is of special interest, particularly in relation to the history of systematic entomology:

My friends Mr Banks and Dr Solander had returned from their voyage round the world (with Captain Cook), and had brought with them innumerable specimens of natural history and insects. I now lived very pleasantly. With Banks, Hunter and Drury I found plenty of objects to engage my time, and everything which could possibly be of service to me. My situation was not only very delightful, but it afforded the means of gaining much instruction. In 1775...my *Systema Entomologiae* appeared. Entomology was at that period in its infancy. We had then only the Systema of Linnaeus, whose classification, derived from the wings of insects, was not the

---

[1] "The Autobiography of John Christian Fabricius," translated from the Danish by the Rev. F. W. Hope, A.M., F.R.S. (*Transactions of the Entomological Society of London*, Vol. IV, 1845).

most natural, and his species were very imperfectly defined. More-over, it contained but few species, as the great founder of that system was fully aware that the science would make little progress by a compilation of an inadequate number of species; the amount of genera also described by him was not great. In my System I made use of the organs of manducation as marks of distinction for my classes and species, and in spite of all its faults, which arose from the smallness of those parts, my classes were far more natural, my species were more numerous and more ably defined, and the number of described genera considerably greater. I at the same time extended the Orismology, fixed its significations with greater accuracy, and introduced the concise language of the Linnaean school in this department of natural history.

Fabricius worked through Dr Hunter's Cabinets of Insects, he identified and labelled the various specimens, and a large number of them were named and described by him (in his *Systema Entomologiae* and his later works) as species new to Science.

It has long been desired that the Fabrician Types in the Hunterian Collection should be made accessible, for purposes of systematic entomology, by the publication of accurate up-to-date descriptions together with accurate figures. When recently, following on the magnificent gift of the Bishop Collection, the University Entomological Collections came under my supervision, it was then suggested by Professor Graham Kerr that I should at once begin this work on the Fabrician types. I have been enabled to do that by the *Carnegie Trustees for the Universities of Scotland*, and I here record my obligations.

The most important part of the work of investigating these types, checking their identity and carefully comparing with them representative modern examples of the species, has been done in London, in the Entomology Department of the British Museum of Natural History, and in close collabora-tion with Dr C. J. Gahan, Mr Gilbert J. Arrow, Mr Kenneth G. Blair and Mr Bryant. To these gentlemen and to Sir Guy A. K. Marshall, C.M.G., F.R.S., Director of the Imperial Bureau of Entomology, I am indebted for their unfailing courtesy and kindness to me and for constant help generously given. Without the advantage of their special knowledge and expert advice, the work could not have been properly done, nor would it have carried the same weight as a contribution to systematic entomology.

For the facilities granted to me while at the British Museum, and especially for permission to consult freely the officers of the Entomology Department, I beg to tender thanks to Sir Sidney Harmer, K.B.E., F.R.S.

I desire to acknowledge my particular indebtedness and express my thanks to the *Carnegie Trustees*, for a special grant towards the cost of production, and to the *Publications Standing Committee of Glasgow University* for similar substantial assistance to ensure the publication of this volume.

I also desire to acknowledge gratefully my indebtedness to Professor Graham Kerr, F.R.S., for his constant interest and for many helpful suggestions.

The figures of the types have been drawn by Miss Margaret Rankin Wilson, D.A. (G.S.A.), and are good examples of her skill as an artist and her accuracy in rendering important detail. My special thanks are due to her for the care and interest she has taken in the work.

The Hunterian Collection of Insects is now housed in the Museum of the new Zoology Department; it is contained in five cabinets herein referred to as Cabinets A, B, C, D and E. The specimens in these cabinets are more than one hundred and fifty years old, and the great majority of them are still in a remarkably good state of preservation; they are arranged in the order given by Fabricius in his *Species Insectorum* (1781). That applies to the species of each genus, but not invariably to the arrangement of the genera. The labels are not attached to the specimens, but are fastened to the bottom of the drawer, and are placed, each label immediately above the specimen or specimens to which it refers.

When Professor Graham Kerr examined the collection, he found that certain drawers (Cabinet A) had been tampered with by some person or persons unskilled; in several in-stances the specimens and labels were obviously misplaced and the labels had been gummed on to the bottom of the drawer, the original pins having been removed. Fortunately I have been able (after considerable trouble and deplorable loss of time) to locate most of the misplaced types; I have not removed these from the drawers in which they were found, but have there indicated them with special labels, and have thus refrained from perpetuating confusion by inter-fering in any way with the existing arrangement of the specimens.

Beginning with the *Coleoptera* I have thus far investi-gated the types of species belonging to that order. My notes on types of the Families *Cicindelidae, Carabidae, Dytiscidae, Scarabaeidae, Silphidae, Histeridae* and *Erotylidae*, form the subject matter of this volume, the first portion of a contri-bution which, it is hoped, may ultimately take the form of a complete Descriptive Catalogue of the Insect Types in the Hunterian Collection at Glasgow University.

In Dr Hunter's Collection there are several specimens which are the types of species founded by the noted French entomologist, Antoine G. Olivier (b. Jan. 17, 1756; d. Nov. 1, 1814), and there are also certain other types supposed or known to have been acquired by Dr Hunter. As it seemed advisable to publish these I have therefore included them.

In the Cabinets each species is represented usually by two specimens; but in several instances there is only one speci-men, or more than two, under the name label, and sometimes the specimens are of different species. When there is no clear indication that one or other of the specimens is to be regarded as the type, I have described as such that one which answers to the descriptions and which corresponds most closely with modern examples of the species in the British Museum Collec-tions.

The descriptions of the types are necessarily limited to those features or characters which I have been able to make out clearly; the mouth appendages and other parts of many of the types are defective or wanting, or are obscured by the accumulated dust of years, and for the best reasons I have not attempted to clean them. The references to the species are those given in the Catalogues.

The coloured plates represent the specimens with all their existing defects due to the age of the Hunterian Collection. No doubt it would be more in agreement with custom to figure these insects with their appendages symmetrically arranged; it was, however, clearly out of the question to subject such historically important types to the risks of relaxing and resetting, and it appeared equally out of the question to restore their imperfections in the drawings. To try to improve their appearance in these ways would certainly lessen the value of the illustrations and serve no useful purpose; the aim has been to portray them exactly as they are at the present time.

The "Bishop" Collection, its wealth of material British and Exotic, has been of great service to me while investigating these Hunterian insects; I have often had occasion to refer to it for modern examples with which to compare the types.

ROBERT A. STAIG

THE ZOOLOGY DEPARTMENT
THE UNIVERSITY, GLASGOW
June, 1930

Order COLEOPTERA

Family CICINDELIDAE

The following are the species of Cicindelidae mentioned by Fabricius, in his published works, as having been described by him from specimens in Dr Hunter's Collection:

Cicindela unipunctata   *Syst. Ent.* p. 225, No. 8 (1775).
   biramosa   *Mant. Ins.* 1, p. 186, No. 20 (1787).
   8-guttata   *Ibid.* p. 187, No. 24.
   cincta[1]   *Ent. Syst.* 1, 1, p. 175, No. 27 (1792).

The above names are the original names as given by Fabricius, and the references are to the works in which these species were first described.

1. *Cicindela biramosa* Fab.

Coleopterorum Catalogus, pars 86 (W. Horn, 1926), Carabidae, Cicindelidae, p. 190. *Fauna of British India*, Coleoptera, Introduction and Cicindelidae (W. W. Fowler, 1912), p. 431.

Ceylon and Southern India, Bengal, etc., Burmah, China.

SYN. *Cicindela biramosa* Fab., *Sp. Ins.* 1, p. 286, No. 16 (1781); *Mant. Ins.* 1, p. 186, No. 20 (1787); *Ent. Syst.* 1, 1, p. 175, No. 28 (1792); *Syst. Eleuth.* 1, p. 240, No. 42 (1801); Oliv. *Ent.* 11, 33, p. 26, pl. 2, fig. 16 *a* and *b*, and pl. 3, fig. 29 (1790).
C. tridentata Thunb., *Nov. Ins. Sp.* p. 26, fig. 40 (1781).

There are two specimens of this species under label

'*Cic. biramosa*
Fabr. pag. 286, No. 16.'

in Cabinet B, drawer 9, which answer the descriptions given by Fabricius and Olivier and (recently) by W. W. Fowler. I have compared them with modern examples of *biramosa* in the British Museum, and also with examples in the 'Bishop' Collection, and they closely agree.

---

[1] Previously described by Olivier, *Ent.* 11, 33, p. 10, pl. 3, fig. 33 (1790).

This species was first described by Fabricius (*Sp. Ins.* 1) from a specimen in the Hattorf Collection, and the habitat is there given as Germany, evidently an error. Later, however, Fabricius (*Mant. Ins.* 1) refers the type to Hunter's Collection and gives India as the habitat. The Hunterian specimens must therefore meanwhile be regarded as doubtfully the co-types.

In the descriptions of *biramosa* the elytral border is stated to be white; but in the Hunterian specimens it is distinctly yellow.

W. W. Fowler mentions that the Ceylon variety (for which he proposed the name *dilatata*) has the white markings much developed; and examination of a fine series of Ceylon specimens in the 'Bishop' Collection shows that the yellow border tends to become markedly thickened from the middle to the apex. One of the 'Bishop' examples, labelled 'South Ceylon', has the light yellow border less thickened and varies little from the Hunterian specimens.

Description of Co-type, *Cicindela biramosa* **Fab.** Moderately large. Form robust. Smooth (glabrous) and glossy greenish black above, the head and the prothorax a little coppery, the elytra with light yellow and irregular biramous or triramous outer borders; the underparts of the head and the prothorax, and also the legs, bright coppery with a thin pubescence of whitish hair; the abdomen brilliant deep violet-blue.

The *head* is narrower than the prothorax; the vertex is convex and rugulose, and has two depressions, one at the side of each eye; the frons is convex; the forehead is marked by concentric semicircular striae around the large and prominent green eyes; the clypeus, which extends laterally in front of the insertions of the antennae, is very narrow; the labrum and the bases of the mandibles are testaceous.

The *pronotum* is oblong and convex; in front it is almost straight, and the sides are rounded in at the base, which is sinuate, with a slight and broad median lobe in front of the

PLATE I

Cicindela biramosa Fab. × 5½
and apical portion of elytron

scutellum. Across the front of the pronotum there is a deeply sunk transverse line, near the base there is a similar line, and between these the central longitudinal line is well marked. The surface of the pronotum is rugulose, rather rugose at the front and the base and about the central line.

The *scutellum* is small, broadly and sharply triangular, raised in the centre, and greenish.

The *elytra* are broader than the prothorax, plano-convex, and parallel-sided; the bases are rounded, the outer sides are a little deflexed, and the apices are obliquely rounded; the inner and outer borders are marginate, but the apical edges are finely serrate, and there is a very small sharp spine at each sutural (apical) angle; the shoulders are well marked, within each there is an elongate impression marked by nine large round punctures; there is also a regular longitudinal row of eight or nine large round punctures, at varying intervals apart, on each elytron near and parallel with the suture. The surface of the elytra is closely punctulate with crescentic punctules, which become smaller towards the apex; a light yellow border extends right round the outer side of each elytron from the shoulder to the sutural angle, the inner edge of this border is irregular and has at the middle a conspicuous transverse extension (knob-like in outline) reaching half-way across the elytron, and a lesser (double) offshoot at the apex.

The *prosternum* has punctate sides, and some of the punctures bear whitish setae; the *metathoracic episterna* are finely punctulate, and some of the punctules bear whitish hairs.

The *legs*, moderately long and slender, are coppery green, with dark metallic green coxae and trochanters; the posterior coxae are large and have strongly developed trochanters.

Length 12 mm.; breadth (across the elytra) 4 mm.
Hab. Germany and India (Fab.), East Indies (Oliv.).
See Plate 1.

2. *Cicindela cincta* Oliv.

Coleopterorum Catalogus, pars 86 (W. Horn, 1926), Carabidae, Cicindelidae, p. 154. Oliv. *Ent.* ii, 33, p. 10, pl. 3, fig. 33 (1790).
Africa.

SYN. *Cicindela cincta* Fab., *Ent. Syst.* i, i, p. 175, No. 27 (1792); *Syst. Eleuth.* i, p. 240, No. 40 (1801).

The specimen in Cabinet B, drawer 9, under label

'*C. cincta*, Oliv.'

(the handwriting presumably Olivier's) corresponds with the descriptions of this species given by Olivier and Fabricius. Mr K. G. Blair has carefully compared it with the modern examples of *cincta* in the British Museum, and he considers it to be the type.

I have compared this type with a series of fifteen modern examples (fourteen from Sierra Leone and one from Ashanti) in the 'Bishop' Collection. The stripes and spots on these 'Bishop' examples are not white, as stated by Olivier and Fabricius, but are (like those on the type) distinctly yellow; only two of the examples have these markings cream-coloured. The spotting is very variable: the Ashanti specimen is without any trace of spots on the elytra. The specimens from Sierra Leone show considerable variation in the size and number of the spots, one having only the two hindmost spots and these greatly reduced, and others being without the middle spots; one specimen is light green above, and two have the head and pronotum light coppery green.

Description of Type, *Cicindela cincta* Oliv. Large. Form robust. Dull greenish black above, the head and the prothorax a little coppery, the elytra metallic violet along the sides and with a narrow dorso-lateral yellowish stripe and three small and unequal yellowish spots on each disc; the underside glossy violet-blue and with a scanty pubescence of short light yellowish hairs, the lower segments of the palps

PLATE 2

Cicindela cincta Oliv. × 4

light red, the thoracic sterna with bright coppery touches, and the upper parts of the legs coppery.

The *head* is longer than broad, its breadth is about the same as that of the prothorax, and its surface is shagreened; the vertex is convex; the frons is flattened between the eyes and is convex and vertical in front, it has prominent angular and coppery supra-orbital ridges, and it is marked by irregularly semicircular striae about the large and prominent eyes; the clypeus, which extends laterally in front of the insertions of the antennae, is very narrow and is violet-coloured; the labrum is light red on the base, the front part is black with a few long setae and with seven teeth on its front edge; the sides of the head are coppery and rugulose, the genae are glossy violet and are closely striated; the base and the first segment of the antennae are coppery, the next three segments are glossy violet and the succeeding segments are light red; the bases of the mandibles and the basal segments of the maxillary and labial palps are light red.

The *pronotum* is a little longer than broad, it is almost straight in front, the sides are rounded and bear light yellowish setae, and the base is sinuate with a slight and broad median lobe in front of the scutellum. Across the front part of the pronotum there is a deeply sunk triangular line, and posteriorly there is a similar but bicrescentic transverse line. These two lines are continuous round the sides of the prothorax and thus form strongly marked constrictions; and the central longitudinal line between them is distinct but short. The front part of the pronotum is convex, two prominent and contiguous rounded elevations form the central area, and the basal part is flat; the pronotal surface is shagreened.

The *scutellum* is small, broadly and sharply triangular, and rugulose.

The *elytra* are broader than the pronotum, plano-convex, and almost parallel-sided; the bases are sinuate, the outer sides are deflexed, and the apices are obliquely rounded; the outer borders are marginate and there is a very small sharp spine at each sutural (apical) angle. The surface of the elytra is punctulate and dull greenish black, coppery violet along the sides, with a narrow and yellowish dorso-lateral stripe extending from each shoulder to the middle of the apical margin, and with three yellowish spots situated well apart and about the middle of each elytron (the foremost spot small and round, the middle one elongate-oval and near the suture, the hindmost and largest one roughly triangular); there are a few light yellowish setae about the shoulders.

The sides of the *prosternum* are rugulose; the *metathoracic episterna* are rugulose and coppery violet; each side of the *metasternum* is punctate and setose and has a brilliant coppery coloured triangular patch.

The *legs*, moderately long and slender, are coppery green; the posterior coxae are large and have strongly developed trochanters.

Length 18 mm.; breadth (across the elytra) 6 mm.
Hab. Central Africa (Oliv.), Africa (Fab.).
See Plate 2.

3. *Cicindela octoguttata* Fab.

Coleopterorum Catalogus, pars 86 (W. Horn, 1926), Carabidae, Cicindelidae, p. 139.

Senegal, etc.

Syn. *Cicindela 8-guttata* Fab., *Mant. Ins.* I, p. 187, No. 24 (1787); *Ent. Syst.* I, 1, p. 177, No. 32 (1792); *Syst. Eleuth.* I, p. 242, No. 51 (1801); Oliv. *Ent.* II, 33, p. 28, pl. 3, fig. 32 (1790).

In Cabinet B, drawer 9, under label

'*Cic. 8-guttata*
Fabr. MSS'

there are four specimens which correspond more or less closely with the descriptions given by Fabricius and Olivier.

Mr Blair and I have compared two of the specimens with modern examples of this species in the British Museum. I

PLATE 3

Cicindela 8-guttata Fab. × 8

have here described the better marked one, which is a closer match than the other and which may be regarded *provisionally* as the type.

Description of Co-type, *Cicindela* 8-*guttata* **Fab.**

Small size. Bronze-green above, with five small whitish spots and two whitish linear marks; glossy blue and coppery and thinly pubescent beneath.

The *head* and the prothorax are equal in width; the vertex is convex behind and shagreened, and flattened in front; the frons is flattened; the forehead is marked by concentric semicircular striae about the large and prominent eyes; the narrow clypeus extends laterally in front of the insertions of the antennae; the labrum and the bases of the mandibles are testaceous.

The *prothorax* is almost cylindrical; the *pronotum* is longer than broad (sub-quadrate), it is rounded in front, the sides are gently rounded, and the base is sinuate, with a small median lobe in front of the scutellum; there is a transverse constriction in front and behind, the central line is obsolete, the surface is finely rugulose and has a shagreened appearance, and about the sides there are a few light-coloured setae.

The *scutellum* is triangular, sharply pointed, with rounded sides and a concave base, and is apically depressed.

The *elytra* are much broader than the thorax, planoconvex, marginate (except at the apices), finely shagreened, strongly punctate, parallel-sided, the outer sides a little deflexed; the apices are finely serrate and are oblique, from the outer angle each elytron slopes downwards and inwards to the suture and is rounded in to the sutural (apical) angle, which bears a very small sharp spine; the shoulders are well marked, and within each shoulder there is a small impression. Upon each elytron there are five whitish rounded spots and two whitish linear marks, and these are situated as follows: one large oval spot on the shoulder, two round spots (one very small) anteriorly, another beyond the middle and near the suture, and the fifth spot towards the outer angle of the apex; about the middle of the outer side there is a whitish transverse and bracket-like mark connected with a whitish linear mark on the margin; the other whitish linear mark extends along the apical margin.

The sides of the *prosternum* are punctate, and the punctures bear long whitish setae. The *metathoracic episterna* are broad and are thinly covered with recumbent light-coloured hairs.

The *legs* are very long and slender; the posterior coxae are large and have strongly developed trochanters.

Length 8 mm.; breadth (across the middle of the elytra) 3 mm. Hab. America (Fab.), South America (Oliv.).
See Plate 3.

4. *Cicindela unipunctata* Fab.

Coleopterorum Catalogus, pars 86 (W. Horn, 1926), Carabidae, Cicindelidae, p. 291. *Catalogue of the Coleoptera of America, North of Mexico* (Charles W. Leng, 1920), p. 42.

N.Y., Georgia, Missouri, Kentucky, Indiana.

SYN. *Cicindela unipunctata* Fab., *Syst. Ent.* p. 225, No. 8 (1775); *Sp. Ins.* 1, p. 285, No. 13 (1781); *Mant. Ins.* 1, p. 186, No. 17 (1787); *Ent. Syst.* 1, 1, p. 174, No. 23 (1792); *Syst. Eleuth.* 1, p. 238, No. 33 (1801); Oliv. *Ent.* 11, 33, p. 23, pl. 3, fig. 27 (1790).

The specimen under label

'*Cic. unipunctata*
Fabr. pag. 285, No. 13'

in Cabinet B, drawer 9, answers to the descriptions given by Fabricius and Olivier; and it also closely agrees with modern examples of this species in the British Museum.

As stated by Fabricius, this species was founded on a specimen in Drury's Collection. Olivier had access to Drury's Collection; but Olivier in his description of *unipunctata* Fab. refers the specimen to Hunter's Collection. Probably this was one of the Drury specimens which were acquired by Hunter.

Description of Type, *Cicindela unipunctata* **Fab.**

PLATE 4

Cicindela unipunctata Fab. × 4

Moderately large. Form robust. Dull bronze-green above, with a roughly triangular cream-coloured spot at the middle of the outer border of each elytron; glossy deep violet-blue beneath, the labrum and the bases of the mandibles testaceous.

The *head* is oblong, its breadth is about equal to that of the prothorax; the vertex is convex and rugulose; the frons is depressed between the eyes, convex and vertical in front, and is marked by longitudinal striae, and beside each triangular supra-orbital ridge there are two long and upright setae; the clypeus, which extends laterally in front of the insertions of the antennae, is very narrow; the labrum is transverse and testaceous, it has three large punctures bearing setae and situated near the front edge which is darkly coloured and tridentate; the sides of the head and the genae are striolate; the bases of the mandibles are testaceous; the antennae, the maxillary palps and the labial palps are coppery green.

The *pronotum* is a little longer than broad, it is nearly straight in front, the sides are rounded and bear light-coloured setae, the base is sinuate with the median part broad and very slightly lobate. Across the front part of the pronotum there is a deeply impressed triangular line, and posteriorly there is a similar transverse and slightly bicrescentic line. These two lines are continuous round the prothorax, forming well-marked constrictions, and between them the central longitudinal line is distinct. The front part of the pronotum is convex, the central area has the form of two prominent and contiguous elevations, and the basal part is flat; the surface is rugulose, a little shagreened and setose on the elevations.

The small *scutellum* is broadly and sharply triangular.

The *elytra* are broader than the pronotum, plano-convex, and slightly rounded at the sides, which are deflexed at the shoulders; the bases are sinuate, and the apices are obliquely rounded; the outer and the apical borders are marginate, and along the outer border at each shoulder there is a short row of tubercles bearing golden setae; the surface of the elytra is punctate and dull bronze-green, with a roughly triangular cream-coloured spot situated about the middle and close to the outer margin of each elytron.

The sides of the *prosternum* and the *metathoracic episterna* are violet-blue and rugulose; the *metasternum* is violet-blue, a little coppery about the median line, and slightly rugulose.

The *legs* are long and slender; the coxae and the femora are glossy violet-blue, the tibiae and the tarsi are bright copper; the posterior coxae are large and have strongly developed coppery green trochanters.

> Length 15 mm.; breadth (across the elytra) 5 mm.
> Hab. America (Fab.), South America (Oliv.).
> See Plate 4.

Family CARABIDAE

The following are the species of Carabidae mentioned by Fabricius, in his published works, as having been described by him from specimens in Dr Hunter's Collection:

Carabus rufescens	*Sp. Ins.* 1, p. 312, No. 73 (1781).
elevatus	*Mant. Ins.* 1, p. 198, No. 37 (1787).
unicolor	*Ibid.* p. 198, No. 38.
pallipes	*Ibid.* p. 202, No. 86.
ruficollis	*Ibid.* p. 203, No. 91.
turcicus	*Ibid.* No. 96.
truncatellus	*Ibid.* p. 206, No. 123.
Scarites depressus	*Ibid.* No. 1.
marginatus	*Ibid.* No. 2.

The above names are the original names as given by Fabricius, and the references are to the works in which these species were first described.

The following are species described by Olivier from specimens in Dr Hunter's Collection:

Carabus scabrosus	*Ent.* III, 35, p. 17, pl. 7, fig. 83 (1795).
planus	*Ibid.* p. 62, pl. 6, fig. 63.

In the following pages these types are described under their *modern* names and in the order adopted in Gemminger and Harold's *Catalogus Coleopterorum*, 1, Carabidae, 1868.

5. *Procerus scabrosus* (Oliv.)

Catalogus Coleopterorum (Gemminger and Harold, 1868), 1, Carabidae, p. 56.

Turkey.

SYN. *Carabus scabrosus* Oliv., *Ent.* III, 35, p. 17, pl. 7, fig. 83 (1795); Fab. *Syst. Eleuth.* 1, p. 168, No. 1 (1801).
C. gigas Creutz., *Ins.* 107, 1, tab. 2, fig. 13.

One specimen in Cabinet B, drawer 10, with an attached label

'*Car. scabrosus*, Oliv.'

(the handwriting presumably Olivier's) and placed under label

'Hab. in urbe Constantinople. Dr Rae'

is evidently the type; the specimen corresponds with the descriptions of the species and it closely agrees with modern examples in the British Museum. I have also compared it with the specimens of *scabrosus* Oliv. in the 'Bishop' Collection.

Olivier in his description has not noted the smaller cusp of the large double tooth of the mandibles.

Description of Type, *Carabus scabrosus* **Oliv.** Large. Elongate-ovate and convex. Deep violet above and very scabrous; the antennae, the legs and the underside of the body dull glossy black, with the sides of the head and thorax, the epipleura and the lateral parts of the abdomen bright violet.

The *head* is large and oblong; the vertex is round, strongly constricted behind the eyes, and across the middle of the vertex there is a broadly triangular transverse ridge behind which the surface is less rugose than in front; the frons is deeply rugose and rather flattened, and it has two anterior depressions; the clypeal suture is represented by a raised sinuate line extending between the anterior ends of the pro-

minent supra-orbital ridges; the clypeus is transverse, almost
straight in front, and it has three depressions, the median one
advanced and smooth; the labrum is transverse and broader
than the clypeus, the sides are very round, the front is
hollowed out, the middle portion is depressed, and the surface
is marked by short longitudinal ridges and impressions and
by a small fovea or pit at each side. The eyes are prominent
and testaceous. The first segment of the antennae (which are
imperfect in this specimen) is stouter than the others, and it
bears a pit near the distal end; the third segment appears to
be the longest. The mandibles are smooth, large and curved,
with sharp hooked tips; and each bears a strong double tooth
or two strong teeth (the larger one curved), a median tooth-
like process on the inner edge, and a close fringe of reddish
hairs.

The *pronotum* is nearly twice the width of the head, and its
breadth is slightly greater than its length; it is broadest
across the middle and is roughly hexagonal, its sides being
angularly rounded; the front is slightly hollowed out and is
narrower than the base, which is sinuate, with a very slight
median lobe; the front angles are sharp, the hind angles are
rounded; the front and the sides are marginate, and the side
margins are upturned; the disc is very convex in front and
somewhat flattened towards the base which is transversely
depressed, the surface of the disc is strongly rugose (scabrous)
and is marked by a median longitudinal impressed line extending from the front margin to the hind margin.

The *scutellum* is broadly triangular, centrally hollowed and
marked with some short longitudinal linear impressions.

The *elytra* are wider than the prothorax, oval, and separately convex so that the suture is distinctly insunk; the bases
are narrow, constricted at the shoulders, rounded, and partly
marginate; the sides are marginate, the outer margins are a
little raised; the surface is very scabrous, but the sculpture
is not irregularly rugose like that of the pronotum, it has the
form of concatenate lobular tubercles.

The *prothoracic episterna* are rugose and violet-coloured. The front *coxal cavities* or *acetabula* are open behind. The *mesothoracic episterna* are punctate and violet-coloured, and the *epimera* reach the middle acetabula. The *metathoracic episterna* are rugose-punctate and violet-coloured. The hindmost *coxae* are not separate and are flattened. The *epipleurae* are mainly smooth and violet-coloured.

The *abdominal sterna* are more or less finely rugulose, with the lateral portions rugulose-punctate and each marked by an irregular shallow depression.

The *legs* are long. The hindmost trochanters are large and cordiform. The femora are grooved ventrally and have parallel rows of punctures bearing short and stiff setae; the hindmost femora are the longest. The tibiae have rows of punctures bearing short spines, and each tibia has two long distal spines; the front tibiae are grooved ventrally and are shorter than the middle tibiae; the hindmost tibiae are twice the length of the front tibiae and are the longest. The tarsi show similar progressive lengthening; the first and longest segments of the tarsi are grooved ventrally, those of the hindmost tarsi have also lateral grooves; the other segments of the tarsi are grooved ventrally; the edges of the grooves are closely set with very short spines.

Length 46 mm.; breadth (across the elytra) 18 mm.
Hab. Constantinople (Oliv.).

6. *Scaphinotus elevatus* (Fab.)

Catalogus Coleopterorum (Gemminger and Harold, 1868), I, Carabidae, p. 84.

North America—Connecticut to Florida.

Syn. *Carabus elevatus* Fab., *Mant. Ins.* I, p. 198, No. 37 (1787); *Ent. Syst.* I, 1, p. 132, No. 33 (1792); Oliv. *Ent.* III, 35, p. 46, pl. 7, fig. 82 (1795).

Two specimens (one defective) under label

'*Car. elevatus*
Fabr. MSS'

in Cabinet B, drawer 11, answer to the descriptions given by Fabricius and Olivier. One of the specimens has been examined and compared by Mr Arrow and Mr Blair, and it closely agrees with the modern examples of *elevatus* in the British Museum.

As pointed out by Mr H. E. Andrewes (*Trans. Ent. Soc. London*, 1919, p. 178), Fabricius subsequently described under this name (*Ent. Syst.* I, 1, p. 162, No. 166, 1792 and *Syst. Eleuth.* I, p. 204, No. 186, 1801) another species, which was found to be identical with Hope's *Coptodera bicincta*, an Oriental insect.

Description of Co-type, *Carabus elevatus* Fab. Form elongate-ovate and convex, with the sides of the pronotum and the elytra considerably elevated. Coppery green above, the pronotum coppery violet and rugose, the elytra irregularly bordered with violet and closely striate-punctate; the underside of the body, the mouth-parts and the legs glossy black, the epipleura of the elytra violaceous with a coppery sheen.

The *head* is coppery green, small, elongate and very narrow; the vertex is round, slightly constricted behind the eyes, and transversely rugulose; the frons is finely punctured, with a few faint foveae along the middle; on the clypeus there are a few short linear impressions, the labrum is elongate and bifurcated, the eyes are prominent and each supra-orbital ridge bears one seta. The first four segments of the antennae are coppery green, the remaining segments are reddish; the first segment is long and stout and it bears a long distal seta, the second and the fourth segments are shorter than the others which are equal in length, and, with the exception of the first two, the segments are pubescent. The mandibles are narrow and prominent. The terminal segments of the palps are dilated and scoop-like.

The *pronotum* is much wider than the head, and its breadth is almost equal to its length; it is widest at the base which is concave and sinuate; the front is widely excavated and is

PLATE 5

Carabus elevatus Fab. × 4

elevated; the sides are greatly produced and elevated and are curved inwards from the middle to the front, the front angles are rounded (very obtuse), the hind angles are almost sharp, and within each hind angle there is a large and smooth light coppery spot. The front and the sides of the pronotum are strongly marginate, the disc is convex and is marked by a median impressed line between a deep transverse crescentic impression in front of the disc and a similar bicrescentic impression behind.

The *elytra*, which are wider than the pronotum, are oval, very convex and apically depressed; the outer borders are strongly marginate and are greatly elevated about the shoulders, which are very prominent and square-cut in form with rounded angles; the surface is strongly and closely striate-punctate, except about the middle and the sides of each elytron where the striae are irregular or obsolete and the surface is rugose-punctate. The epipleura are angular, wide about the base, and rugose-punctate.

The anterior coxal cavities are open behind; the posterior coxae are separated. The abdominal sterna (except the last one) have each two seta-bearing pores, one on each side near the middle line and near the posterior edge. The femora are smooth, the tibiae and the tarsi bear short spine-like hairs; the segments of the front and the middle tarsi are pad-like beneath, and the segments of the hindmost tarsi have short distal tufts of golden hair.

Length 19 mm.; breadth (across the elytra) 9 mm.
Hab. South America (Fab.).
See Plate 5.

7. *Scaphinotus heros* Harris

Catalogue of the Coleoptera of America, North of Mexico (Charles W. Leng, 1920), p. 43.

United States of America.

Syn. *Carabus unicolor* Fab., *Mant. Ins.* I, p. 198, No. 38 (1787); Oliv. *Ent.* III, 35, p. 47, pl. 6, fig. 62 (1795).

The specimen under label

'*Car. unicolor*
Fabr. MSS'

in Cabinet B, drawer 11, answers to the descriptions of *unicolor* given by Fabricius and Olivier.

In the British Museum Collection there is only one example (a female) of *heros*, and it very closely resembles the Hunterian *unicolor*, which is evidently the type.

In Gemminger and Harold's Catalogue *unicolor* Fab. and *heros* Harris are given as distinct species; but according to Casey (*Memoirs*) they are synonymous.

In his description of this species, Fabricius says, "perhaps only a variety of the preceding (*elevatus*)".

Description of Type, *Carabus unicolor* Fab. Form oval, broad and convex, with the sides of the pronotum and the elytra elevated; concolorous, dull glossy black.

The *head* is small, elongate and very narrow, faintly rugulose (almost smooth), and slightly constricted between the vertex and the frons (behind the eyes). On the clypeus there are two small median ridges between which are two small and shallow impressions, one behind the other; the clypeal suture is a finely marked and roughly semicircular line between the prominent supra-orbital ridges. The labrum is bifurcate to the base and it bears six setae (four median and short, two outer and long). The eyes are testaceous and prominent and are without supra-orbital setae. The first four segments of the antennae are glossy black, the remaining segments are reddish; the first segment is long and stout and it bears a long distal seta, the second and the fourth segments are shorter than the others which are of equal length, and, with the exception of the first two which are smooth or bare and the third and fourth which have a few hairs, the segments are pubescent. The mandibles are long and narrow and prominent. The terminal segments of the labial and maxillary palps are greatly dilated.

PLATE 6

Carabus unicolor Fab. × 3

The *pronotum* is much wider than the head; it is transverse and marginate, and is widest at the base, which is very concave and with a distinct median lobe; the front is excavate, almost straight, and a little raised; the sides are incurved, almost semicircular, strongly marginate and elevated, the front and the hind angles are a little rounded; the disc is convex and is marked by a median impressed line between a small transverse crescentic impression in front of the disc and a very deep transverse impression behind. The surface of the pronotum is irregularly and not closely punctate, except the middle of the disc which is smooth.

The *elytra* are oval and broad, nearly twice the breadth of the pronotum, and convex; the outer borders are strongly marginate and upturned, considerably elevated about the prominent shoulders. The surface of the elytra is strongly and closely striate-punctate; but the striae and costae become obsolete as such at the outer sides and towards the apices, which are rugose. The epipleura are angular, widest near the base, and rugose-punctate.

The anterior coxal cavities are open behind; the posterior coxae are separated. The abdominal sterna (except the last one) have each two seta-bearing pores, one on each side near the middle line and near the posterior edge. The femora are smooth, the tibiae and the tarsi bear short spine-like hairs, and the tarsi have also short tufts of golden hair beneath the segments at the distal end.

Length 28 mm.; breadth (across the elytra) 14 mm.
Hab. South America (Fab.).
See Plate 6.

8. *Leistus ferrugineus* (Linn.)

Catalogus Coleopterorum (Gemminger and Harold, 1868), I, Carabidae, p. 54.

SYN. *Carabus rufescens* Fab., *Syst. Ent.* p. 247, No. 58 (1775); *Sp. Ins.* I, p. 312, No. 73 (1781); *Mant. Ins.* I, p. 204, No. 104 (1787); *Ent. Syst.* I, 1, p. 162, No. 169 (1792); *Syst. Eleuth.* I, p. 205, No. 191 (1801); Oliv. *Ent.* III, 35, p. 101, pl. 12, fig. 146 (1795).

This species is represented by two specimens under label

'Car. rufescens
Fabr. pag. 312, No. 73'

in Cabinet B, drawer 11, which closely correspond with the descriptions of *rufescens* given by Fabricius and Olivier. Mr Arrow has examined these specimens and has compared them with examples in the British Museum Collection; and he finds that *Carabus rufescens* is identical with *ferrugineus* Linn. (syn. *spinilabris* Panz.) and is not the insect at present known as *rufescens*. The type of the modern *rufescens* is *praeustus* Fab.

I have examined a series of British examples of *ferrugineus* in the 'Bishop' Collection; the majority of them perfectly resemble the type *rufescens* Fab., but there are some which show considerable variation in colour. A few of the specimens are of a shade of tawny brown darker than that of the type; two are very light-coloured, distinctly fulvous excepting the head which is fuscous, and these light-coloured examples have black eyes. Distinguishing the two species, *ferrugineus* and *rufescens*, W. W. Fowler says that in the former "the striae become shallower at the sides, but are always traceable", while in the latter "the striae are feebler and almost obsolete at the sides". I think that in this respect the difference between them is not substantial. The black head and the blunter angles of the thorax in *rufescens* appear to be the best distinctions.

Description of Co-type, *Carabus rufescens* **Fab.** Form elongate-oval and convex. The head and thorax shining tawny brown (fuscous), the elytra clear light tawny brown, fuscous towards the bases; the underside and the legs reddish brown, darker about the thoracic sterna; the antennae and the mouth-parts (labrum, mandibles and palps) light tawny brown.

The *head* is large, its width is a little less than that of the pronotum, and it is oblong and convex; the vertex is rounded

PLATE 7

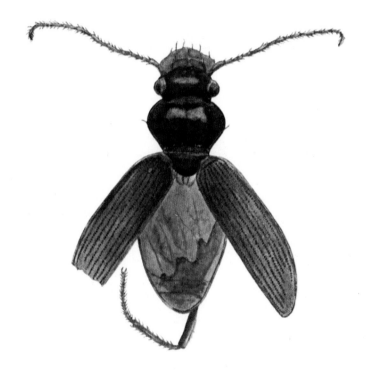

Carabus rufescens Fab. × 12

and is sharply marked off from the frons by a strong constriction behind the large and prominent eyes, each of which
has one long supra-orbital seta; the frons has three or four
very small median foveae; the clypeal suture is a faint line
ending in a pore near the base of each antenna, the front
of the clypeus is straight; the labrum is transverse, semicircular, narrower than the clypeus, and it bears four setae.
The first segment of the antennae is long and stout and it
bears a long distal seta; the second and the fourth segments
are shorter than the third, which is a little shorter than the
first; the other segments are equal in length to the first
segment; the segments are pubescent, excepting the first four.
The mandibles are very broad, with horizontally expanded
sides and slightly hooked tips, and each has two punctures
bearing long setae. The maxillary palps are long and three-
segmented; the first segment is longer and stouter than the
other two, the terminal segment is ovate and truncated at the
tip. The labial palps are shorter than the maxillary palps
and are two-segmented; the first segment is long and thin,
the terminal segment is shorter and clavate.

The *pronotum* is wider than the head, transverse, cordate
in shape and marginate; the front margin is slightly sinuate,
the base margin is straight, the front angles are rounded, the
hind angles are sharp right angles, and the sides are greatly
rounded, narrowing towards the base; the disc is very convex
and smooth, it is divided by a distinct median longitudinal
impressed line, which extends between a transverse semicircular depression at the front and the flattened narrow band-
like base; the frontal depression and the base are punctate,
and at the middle of each side there is a long marginal seta.

The *elytra* are a little wider than the pronotum, oval,
broadest behind the middle, separately convex, and narrowed
at the shoulders; the bases are strongly marginate and the
outer borders are marginate, except the apical portions, which
are also slightly sinuate. The surface of the elytra is marked
with strongly and closely punctured striae, and there is a

distinct scutellary striole; the striae become faint or obsolete about the apical callus and on the outer side.

The *prosternum* is rugose-punctate, with the *episterna* smooth. The *metasternum* is smooth about the middle, elsewhere it is punctate. The abdominal sterna are smooth, except the first three which have the lateral portions punctate. The hindmost coxae are separate, flattened, lobe-like and partly punctate. The hindmost trochanters are large and lobe-like. The hindmost legs are longer than the others, which are about equal in length. The distal ends of the tibiae and the undersides of the tarsi are clothed with a short pubescence of golden hairs.

Length 6 mm.; breadth (across the elytra) 3 mm.
Hab. England (Oliv.).
See Plate 7.

9. *Pasimachus depressus* (Fab.)

Catalogue of the Coleoptera of America, North of Mexico (Charles W. Leng, 1920), p. 47. *Catalogus Coleopterorum* (Gemminger and Harold, 1868), I, Carabidae, p. 176.

North America (New York to Florida).

SYN. *Scarites depressus* Fab., *Mant. Ins.* I, p. 206, No. 1 (1787); *Ent. Syst.* I, 1, p. 94, No. 1 (1792); *Syst. Eleuth.* I, p. 123, No. 1 (1801); Oliv. *Ent.* III, 36, p. 5, pl. 2, fig. 15 (1795).

The specimen under label

'*Scar. depressus*
Fabr. MSS'

in Cabinet B, drawer 12, corresponds perfectly with the descriptions of this species given by Fabricius and Olivier; it also closely matches the modern examples of *depressus* in the British Museum, and is evidently the type.

Fabricius gives Cayenne as the habitat, and Olivier, describing a specimen in the British Museum, also states Cayenne.

For the original description of this species Leng gives *Syst. Ent.*, which is an error.

Description of Type, *Scarites depressus* Fab. Form oval and plano-convex; uniformly glossy black.

The *head* is large, deeply insunk in the thorax (without a distinct neck), transverse (excluding the mandibles), roughly rectangular, smooth and flattened, with two deep median longitudinal depressions on the forehead, and a small oval and oblique slit-like impression at each front angle; the fronto-clypeal suture is faintly indicated across the median depressions to the slits and front angles; the clypeus is slightly and widely excavate in front, the deeply impressed straight line of the excavation strongly marks off the horizontal labrum; the labrum is short and is about half the width of the forehead, and its rugose surface shows a median truncate elevation marked by a short median longitudinal slit, a deep puncture at each outer side, and two or three deeply impressed short longitudinal lines between the elevation and each puncture. The eyes are laterally prominent; the antennae are imperfect in this type specimen; the mandibles are nearly as long as the head, strongly striate, broad, blunt-tipped, and each has a thick and blunt middle tooth which, on the left mandible, is twice as broad as the one on the right and is ridge-like.

The *pronotum* is fully a third wider than the head and is flattened about the middle, a little convex at the sides, transverse, and wider in front than behind; the front is deeply and widely excavate and slightly sinuate, the base is sinuate and marginate; the sides, which are gently rounded and contracted at the base, are expanded as a narrow shelf-like and strongly marginate border, and this border turns within the prominent front angle and merges in the end of a frontal transverse impression; the surface of the disc is smooth, except for the faintly marked median longitudinal line, a slightly wrinkled fovea on each side of the line near the base, and a large puncture at each hind angle; the hind angles are almost sharp, nearly right angles, and the front angles are blunted lobe-like projections.

The *elytra*, which are separated from the thorax by a distinct peduncle, are slightly wider than the pronotum, obovate (wide at the base), dorsally flattened and convex at the sides, which have narrow and marginate borders; the base is broad and sinuate, with rounded humeral angles, and is deeply hollowed out between the prominent shoulder ridges; the surface of the elytra is smooth, except along the inner edge of each side border, from the shoulder angle to the apex, where there is a row of punctures or pores bearing short golden hairs. The epipleura are wide at the base and smooth.

The anterior coxal cavities are closed behind. The meso-thoracic epimera do not reach the middle acetabula. The hindmost coxae are not separated and are flattened, and each has a large puncture or pore. The abdominal sterna are marked by an irregular depression at each side and two punc-tures or pores on the middle portion of the posterior edge. The hindmost trochanters are large and ovoid and each is marked by a large puncture or pore. The front tibiae are broadened towards the distal ends, each is tridentate (in-cluding the terminal 'tooth') and has two long spines (one on the inner edge and one terminal); the middle and the hind tibiae are grooved and have rows of punctures bearing short spines; the middle tibiae have each three long terminal spines and the hind tibiae have two.

Length 28 mm.; breadth (across the elytra) 12 mm.
Hab. Cayenne (Fab. and Oliv.).
See Plate 8.

10. *Pasimachus marginatus* (Fab.)

Catalogue of the Coleoptera of America, North of Mexico (Charles W. Leng, 1920), p. 47. *Catalogus Coleopterorum* (Gemminger and Harold, 1868), 1, Carabidae, p. 177.

North America (South Carolina, Florida, Louisiana).

SYN. *Scarites marginatus* Fab., *Mant. Ins.* 1, p. 206, No. 2 (1787); *Ent. Syst.* 1, 1, p. 94, No. 2 (1792); *Syst. Eleuth.* 1, p. 123, No. 2 (1801); Oliv. *Ent.* III, 36, p. 5, pl. 2, fig. 20 (1795).

PLATE 8

Scarites depressus Fab. × 3

The specimen under label

'*Scar. marginatus*
Fabr. MSS'

in Cabinet B, drawer 12, is the type; it answers the description of this species given by Fabricius and quoted by Olivier, and it closely resembles the British Museum examples with which it has been compared. I have also compared it with a series of modern examples from Florida in the 'Bishop' Collection. Most of the 'Bishop' examples agree with the type; but the red spots on the last abdominal sternum are faint in some and absent in one example.

The locality stated by Fabricius is Cayenne.

For the original description of this species Leng gives *Syst. Ent.*, which is an error.

Description of Type, *Scarites marginatus* Fab.
This species closely resembles the preceding; the differences are printed in small capitals in the following description. Form oval and plano-convex, WITH LIGHTLY FURROWED ELYTRA; glossy black, THE FURROWS OR INTERCOSTAL INTERVALS OF THE ELYTRA DULL BLACK, THE OUTER BORDERS OF THE PRONOTUM AND ELYTRA BLUISH.

The *head* is large, deeply insunk in the thorax (without a distinct neck), transverse (excluding the mandibles), roughly rectangular, smooth and flattened, with two deep median longitudinal depressions on the forehead; THE FRONTO-CLYPEAL SUTURE IS REPRESENTED BY A DEEPLY-IMPRESSED LINE at each side between the median depression and the front angle; the clypeus is slightly and widely excavate in front, the deeply impressed straight line of the excavation strongly marks off the horizontal labrum; the labrum is short and is about half the width of the forehead, and its RUGULOSE surface shows a median truncate elevation and a deep puncture or pore (bearing a long seta) at each outer side, with a few LIGHTLY IMPRESSED short longitudinal lines between the elevation and each puncture. The eyes are laterally prominent.

The antennae are short and are moniliform towards the end; the first antennal segment is long, stout and glossy black and is marked by a pore which bears a long seta; the other segments are short, mostly pubescent and reddish brown. The mandibles are the same length as the head, SLIGHTLY STRIATE, broad, blunt-tipped, and each has a thick and blunt middle tooth.

The *pronotum* is about one-third wider than the head, flattened about the middle, a little convex at the sides, transverse, and WIDER BEHIND THAN IN FRONT; the front is deeply and widely excavate and STRAIGHT, the base is STRONGLY SINUATE and marginate; the sides, which are gently rounded and SLIGHTLY CONTRACTED AT THE BASE, are expanded as a narrow shelf-like and strongly marginate border, and, at the prominent front angle, this border merges in the end of a frontal transverse impression; the surface of the disc is FINELY STRIGOSE, NEAR THE BASE THERE ARE TWO DEEP CRESCENTIC DEPRESSIONS, ONE AT EACH SIDE OF THE MEDIAN LINE, AND ACROSS THE MEDIAN LINE THERE IS A DEEPLY MARKED TRANSVERSE LINEAR DEPRESSION AND SEVERAL PARALLEL TRANSVERSE LINES, and at each hind angle there is a large puncture; the hind angles are SHARP, the front angles are blunted lobe-like projections.

The *elytra*, which are separated from the thorax by a distinct peduncle, are slightly wider than the pronotum, obovate (wide at the base), dorsally flattened and convex at the sides, which have narrow and marginate borders; the base is broad and sinuate, with rounded humeral angles, and is deeply hollowed out between the prominent shoulder ridges; THE SURFACE OF THE ELYTRA IS MARKED BY LIGHT FURROWS AND PROMINENT PARALLEL COSTAL RIDGES WHICH UNITE TOWARDS THE APEX, and along the inner edge of each side border, from the shoulder angle to the apex, there is a row of punctures or pores which bear short golden hairs. The epipleura are wide at the base and smooth.

The anterior coxal cavities are closed behind. The hind-

PLATE 9

Scarites marginatus Fab. × 3

most coxae are not separated and are flattened, and each has a large puncture or pore. The abdominal sterna are smooth (faintly strigose), with THE LATERAL PORTIONS RUGULOSE; and ON THE LAST STERNUM THERE ARE TWO ANTERO-LATERAL RED SPOTS. The hindmost trochanters are large and ovoid, and each is marked by a large puncture or pore. The front tibiae are broadened towards the distal ends, each is tridentate (including the terminal 'tooth') and has two long spines (one on the inner edge and one terminal); the middle and the hind tibiae are grooved and have rows of punctures bearing short spines; the middle tibiae have each three long terminal spines and the hind tibiae have two.

Length 28 mm.; breadth (across the elytra) 12 mm.
Hab. Cayenne (Fab.).
See Plate 9.

11. *Scarites subterraneus* Fab.

Catalogue of the Coleoptera of America, North of Mexico (Charles W. Leng, 1920), p. 47. *Catalogus Coleopterorum* (Gemminger and Harold, 1868), I, Carabidae, p. 187.

Southern California, U.S.A.

SYN. *Scarites subterraneus* Fab., *Syst. Ent.* p. 249, No. 1 (1775); *Sp. Ins.* I, p. 314, No. 2 (1781); *Mant. Ins.* I, p. 206, No. 4 (1787); *Ent. Syst.* I, 1, p. 95, No. 4 (1792); *Syst. Eleuth.* I, p. 124, No. 8 (1801); Oliv. *Ent.* III, 36, p. 8, pl. 1, fig. 10 (1795).

The type of this species is stated by Fabricius to be in the *Lewin* Collection. Olivier described and figured a specimen in Hunter's Collection. There are two specimens in Cabinet B, drawer 12, under label

'*Scar. subterraneus*
Fabr. pag. 314, No. 2'.

These have been examined, and one has been compared with the British Museum examples of *subterraneus*; but it does not correspond, and we have not been able to match it with any of the determined species in the British Museum Collection.

For the original description of this species Leng gives *Mant. Ins.*, which is an error.

12. *Lebia turcica* (Fab.)

Catalogus Coleopterorum (Gemminger and Harold, 1868), 1, Carabidae, p. 141.

Southern Europe.

SYN. *Carabus turcicus* Fab., *Mant. Ins.* 1, p. 203, No. 96 (1787); *Ent. Syst.* 1, 1, p. 161, No. 161 (1792); *Syst. Eleuth.* 1, p. 203, No. 181 (1801); Oliv. *Ent.* III, 35, p. 98, pl. 6, fig. 68 *a* and *b* (1795).

The specimen under label

'*Car. turcicus*
Fabr. MSS'

in Cabinet B, drawer 11, is apparently the type; it answers the descriptions given by Fabricius and Olivier, and it closely resembles the modern examples of this species in the British Museum Collection.

The locality stated by Fabricius is England, which is probably correct. The British records of this species are apparently few. In the 'Bishop' Collection there are two British specimens labelled

'Coll. E. Saunders'

one from Hastings and the other from Dawlish. The Dawlish specimen has a large triangular red patch upon the apex of each elytron.

Description of Type, *Carabus turcicus* Fab. The *head* is about half the width of the pronotum, elongate, flattened, with a short rounded neck; the vertex and the frons are glossy black, the clypeus and the labrum are light red, the underside and the mouth-parts are light red. The eyes are large and prominent and reddish, and there are two supra-orbital setae. The antennae are long and light red, setose and covered with a fine pubescence, except the first three segments which are smooth; the first segment is longer and stouter than the others.

The *pronotum* is subcordiform and is twice as broad as the head; the front is a little concave, slightly emarginate and broader than the base, which is constricted, flattened and band-like, marginate and slightly convex; the sides are strongly rounded and have expanded shelf-like marginate borders, and each margin bears two long setae, one near the middle and the other on the hind angle; the front angles are rounded, the hind angles are sharp right angles; the disc of the pronotum is convex, glossy light red, finely wrinkled and divided by a well-marked median line. The *scutellum* is red.

The *elytra*, which are black with a large irregular light red patch about the shoulders and with red outer margins, are about twice the breadth of the pronotum, plano-convex, parallel-sided and marginate; the bases are rounded with a small sinuation near the suture, the apices are obliquely truncate and sinuate; the surface of the elytra is striate, the striae become looped and united towards the apex, the striae and the slightly convex interstices are very faintly punctured, and there is a distinct scutellary striole between stria 1 and the suture.

The thoracic sterna are light red. The abdominal sterna are light red about the middle and have the lateral parts black.

The *legs* are light red; the inner edges of the front tibiae are deeply notched near the distal end, the penultimate segments of the tarsi are emarginate, and the tarsal claws are pectinate.

Length 5 mm.; breadth (across the elytra) 3 mm.
Hab. England (Fab.), France, Italy (Oliv.).

13. *Metabletus truncatellus* (Linn.)

Catalogus Coleopterorum (Gemminger and Harold, 1868), 1, Carabidae, p. 133.

Europe.

SYN. *Carabus truncatellus* Linn., *Fn. Suec. Nr.* 814 (1746).

Professor Graham Kerr has included this species in his

preliminary notice of the Fabrician Types in Dr Hunter's
Collection; but the type is pre-Fabrician, and there is no
reference to Hunter's Collection in any of the descriptions of
truncatellus by Fabricius and Olivier.

The species is represented by three specimens under label

'*Car. truncatellus*
Fabr. MSS'

in Cabinet B, drawer 11. One of these specimens is not
truncatellus; the other two are very defective and both are
without heads, but their remaining parts correspond with
the original descriptions. I have also compared them with
modern examples in the British Museum and in the 'Bishop'
Collection, and I have been able to establish their identity
with *truncatellus* Linn.

14. *Callida ruficollis* (Fab.)

Catalogus Coleopterorum (Gemminger and Harold, 1868), 1,
Carabidae, p. 116.

Natal. East to West Africa.

SYN. *Carabus ruficollis* Fab., *Mant. Ins.* 1, p. 203, No. 91 (1787);
Ent. Syst. 1, 1, p. 139, No. 65 (1792); *Syst. Eleuth.* 1, p. 185,
No. 80 (1801); Oliv. *Ent.* 111, 35, p. 93, pl. 7, fig. 78 (1795).

The specimen under label

'*Car. ruficollis*
Fabr. MSS'

in Cabinet B, drawer 11, is evidently the type; it corresponds
exactly to the descriptions given by Fabricius and Olivier,
and it closely resembles the modern examples of this species
in the British Museum Collection.

Fabricius, in his earliest description, and Olivier give
South America as the habitat. Later (in *Syst. Eleuth.* 1,
p. 185, No. 80) Fabricius states Guinea and refers to a specimen in "Mus. D. de Schestedt".

The only Fabrician reference given by Gemminger and
Harold is *Syst. Eleuth.*, and the locality stated is Sierra Leone.

Description of Type, *Carabus ruficollis* **Fab.** Form elongate and narrow; the head mainly greenish black; the neck, the eyes, the two basal joints of the antennae, the tips of the mandibles, the thorax, the upper portions of the femora, and the end of the abdomen light red (ferrugineous); the elytra dark metallic bluish green, with a suffusion of light red, and the abdomen (except the end) glossy black.

The *head*, with a distinct neck and with large and prominent eyes, is almost as wide as the pronotum, and is oblong, plano-convex, greenish black, and punctate, with some irregular raised lines along the lateral parts of the forehead; the supra-orbital setae are wanting, but there are two long setae on the vertex; the fronto-clypeal suture is a well-marked crescentic line; the clypeus and the labrum are transverse and smooth. The mandibles are broad, with the bases pitchy and the tips light red and sharply hooked. The antennae are moderately long, filiform, and fuscous (except the first two segments, which are light red); the first or basal segment is the longest and stoutest, the second segment is the shortest, the third is almost as long as the first, the fourth and the fifth are progressively shorter, and the succeeding segments are all about equal in length; all the segments bear setae and (excepting the first two, which are glabrous) they are clothed with a fine pubescence.

The *pronotum* is slightly wider than the head, elongate, broader in front than behind (Fabricius says obcordate), convex with flattened borders, and marginate; the front and the base are straight, the sides are rounded and are contracted about the base; the front angles are rounded and the hind angles are a little rounded, almost right angles; the disc is punctate and is divided by the deeply impressed median line which extends between the transverse linear impressions of the front and hind borders. The *scutellum* is light red and triangular.

The *elytra*, which are separated from the thorax by a distinct peduncle, are dark metallic bluish green, about one-third wider than the pronotum, plano-convex, with the apices truncate and a little wider than the bases, and marginate except at the base; the outer margins are rounded, with a slight but distinct sinuation below the very rounded shoulders; the surface of the elytra is striate-punctate, the slightly convex interstices are less closely punctured than the striae, the striae become looped and united at the apices, and there is a scutellary striole between stria 1 and the suture; on each outer margin near the shoulder there is a row of long setae. The epipleura are wide at the shoulders, metallic blue and punctate.

The upper parts of the femora are light red; the lower parts of the femora and the upper parts of the middle and hind tibiae are black. The front tibiae are black; the inner edges are deeply notched near the distal end, and the two terminal spines are longer than those of the other legs. The tarsi are reddish black; the penultimate segments are bilobed and the claws are strongly pectinate.

Length 8½ mm.; breadth (across the elytra) 3½ mm.
Hab. South America (Fab. and Oliv.), Guinea (Fab. 1801).

15. *Pheropsophus aequinoctialis* (Linn.)

Amoen. Acad. vi, p. 395 (1763). *Catalogus Coleopterorum* (Gemminger and Harold, 1868), 1, Carabidae, p. 102.

Tropical America (Brazil, etc.).

SYN. *Carabus complanatus* Fab., *Syst. Ent.* p. 242, No. 33 (1775); *Sp. Ins.* 1, p. 306, No. 41 (1781); *Mant. Ins.* 1, p. 200, No. 57 (1787); *Ent. Syst.* 1, 1, p. 144, No. 90 (1792).
Brachinus complanatus Fab., *Syst. Eleuth.* 1, p. 217, No. 2 (1801).
Carabus planus Oliv., *Ent.* iii, 35, p. 62, pl. 6, fig. 63 (1795).

The type of this species is pre-Fabrician. The specimen under label

'*Car. complanatus*
Fabr. pag. 306, No. 41'

in Cabinet B, drawer 11, answers the descriptions given by Fabricius.

Olivier described this species under the name *planus* from a specimen in Hunter's Collection. Olivier's figure represents the insect with closed elytra, whereas the Hunterian specimen has the elytra and the functional wings spread out; probably, however, the Hunterian specimen is the Olivier type.

The habitat given by Fabricius and Olivier is St Domingo.

In Gemminger and Harold's Catalogue the only Fabrician reference given for *complanatus* is *Syst. Eleuth.*; but the description in *Syst. Eleuth.* is merely a quotation of his earlier descriptions.

Description of Type, *Carabus planus* Oliv. Reddish and testaceous above, with two broad and irregular black bands (one interrupted) across the lightly furrowed elytra; the underparts glossy pale yellow.

The *head*, without a distinct neck and with moderately large and prominent pale green eyes, is nearly as wide as the pronotum, oblong, dull pale red and mainly smooth; the vertex is convex, the forehead is depressed and is marked by two wrinkled foveae alongside the eyes and two deep longitudinal furrows continuous from the foveae to the clypeus; the supra-orbital ridges are thickened and each bears one posterior seta; the fronto-clypeal suture is straight; the clypeus is transverse and is concave in front, it is marked with small depressions and it bears two setae; the labrum is shorter and a little narrower than the clypeus, straight in front, and on its front border there are five punctures or pores with long setae. The mandibles are broad, with black and blunted tips, and each bears a long seta. The outer lobes of the maxillae are two-jointed, the four-jointed maxillary palps have truncate tips; the mentum is angular-sided, hollowed out and pitted; the ligula is emarginate, the paraglossae are setose, and the labial palps are three-jointed with truncate tips. The antennae are long, filiform, testaceous, finely pubescent and setose at the distal ends of the segments; the first segment is distinctly stouter than the others and less pubescent, the second segment is the shortest, the third segment is the longest, and the length of the succeeding segments is approximately that of the first.

The *pronotum* is a little wider than the head, subcordiform, dull pale red, slightly broader in front than behind, convex with the front and the base slightly flattened and a little rugulose; the front and the base are almost straight and are fringed with very short golden hairs; the sides are very sinuate, rounded out towards the middle and strongly incurved or contracted before the base, and are strongly marginate; the front angles are sharp, the hind angles are a little rounded and projecting; the impressed median line nearly reaches the front and hind margins; the surface of the disc is very faintly wrinkled, and there are a few large pores bearing long setae on the front and hind borders, also two near the middle of each outer margin. The *scutellum* is pale red, and is broadly and sharply triangular.

The *elytra* are about twice the breadth of the thorax, planoconvex, and marginate; the apices are truncate, wider than the narrowed bases, and fringed with very short light-coloured hair; the outer margins are gently rounded, the shoulders are rounded and prominent. The elytra are testaceous with black inner margins and two broad but irregular black bands, one beyond the middle and an interrupted one across the shoulder region; the surface is regularly costate and furrowed, the costae mostly extend to the apical margins and the furrows are faintly wrinkled. The epipleura are wide about the shoulders and roughly angular at the middle of the lower edges. The wings are well developed.

The front coxal cavities are closed behind. The mesothoracic epimera do not reach the middle coxal cavities, which are closed by the sterna. The hind coxae are not completely separated. The underside of the body and the legs are glossy pale yellow and very lightly clothed with short and fine hairs. The hind trochanters are large and ovate. Each front tibia has a deep semicircular notch on the inner edge near the distal end; behind the notch is a long black spine and in front of the notch there is a stouter terminal one. The middle tibiae have each one long terminal spine, the hind tibiae have two. The first segment of the middle and hind tarsi is twice the length of the succeeding segments. The claws are simple.

Length 17½ mm.; breadth (across the elytra) 8 mm.
Hab. St Domingo (Fab. and Oliv.).

16. *Agonoderus pallipes* (Fab.)

Catalogue of the Coleoptera of America, North of Mexico (Charles W. Leng, 1920), p. 75. *Catalogus Coleopterorum* (Gemminger and Harold, 1868), I, Carabidae, p. 250.

North Carolina to Texas, Arizona, Indiana.

Syn. *Carabus pallipes* Fab., *Mant. Ins.* I, p. 202, No. 86 (1787); *Ent. Syst.* I, 1, p. 159, No. 151 (1792); *Syst. Eleuth.* I, p. 200, No. 165 (1801); Oliv. *Ent.* III, 35, p. 89, pl. 9, fig. 99 (1795).
C. americanus Dej.

There are two specimens of this species under label

'*Car. pallipes*
Fabr. MSS'

in Cabinet B, drawer 11. The smaller specimen has one elytron wanting, and the larger specimen is defective also, being without antennae; they correspond with Olivier's figure, but the larger one answers most nearly the descriptions given by Fabricius and Olivier.

The citation of the original description is given by Gemminger and Harold, and also by Leng, as *Ent. Syst.*; but it should be *Mant. Ins.*

Description of Type, *Carabus pallipes* **Fab.** Brownish black above; the pronotum broadly bordered with lighter brown; the elytra with the outer half, the apex, and the distal part of the inner margin reddish testaceous. The first or basal joint of the antennae light red. The underside of the head and the abdomen glossy black, the thorax brownish black, and the legs light red.

The *head*, without a distinct neck and not constricted behind, is oblong, moderately convex and mainly smooth; on the forehead there are two foveae, a deeply impressed semi-circular line passes from the middle of each supra-orbital ridge to each fovea, and between the foveae is the well-marked fronto-clypeal suture. Each supra-orbital ridge bears two setae (the posterior one is longer than the other). The transverse clypeus is concave in front, and is there marked by a transverse depression and four short furrows. The labrum is narrowed, its length is nearly equal to its breadth, its straight front edge bears six long setae and there are a few short setae on each side. The antennae of this specimen are almost entirely wanting, the only remaining part being the light red basal segment of the left antenna.

The *pronotum* is wider than the head and is orbicular, with the sides very strongly rounded; the front is wider than the base and is hollowed out; the base is slightly sinuate, with a wide but very slight median lobe; the disc, which is dark brown with a broad border of lighter brown, is moderately convex, very smooth and shining, and is marked by two deep depressions (one near each hind angle), where the surface is also punctate. The pronotum is marginate along the sides and right round the angles; the front angles are a little rounded, the hind angles are very strongly rounded; on each side margin there are three long setae (one near the shoulder, one about the middle and one at the hind angle); the faintly impressed median line does not reach the front and hind margins. There is a distinct peduncle between the thorax and the elytra. The *scutellum* is cordate.

The *elytra*, which are broader than the thorax, are broadest about the middle, and are plano-convex and marginate; the bases are sinuate and the shoulders are rounded, the apices are angularly rounded and with a well-marked sinuation at the outer angle. The outer half, the apical portion and also the distal part of the inner margin of each elytron are reddish testaceous; the inner half is brownish black. The surface of the elytra is strongly striate-punctate; the striae converge towards the apex and there unite. The scutellary striole does not unite with stria 1, but is between the suture and stria 1. The interstices are convex and smooth. The epipleura are broad at the shoulder.

The *legs* are light red. The front and middle coxae are round, the hind coxae are flat and are not separated and each bears a large pore. The hindmost trochanters are very long and ovoid. The front tibiae have a deep semicircular notch on the inner edge near the distal end; behind the notch is a long spine, and in front of the notch there is a much stouter terminal spine. The middle and hind tibiae have each two terminal spines. The claws are simple.

Length 8 mm.; breadth (across the elytra) 4 mm.
Hab. America (Fab.).

Description of the *smaller specimen*. Brownish black above, the pronotum with a broad reddish testaceous border, the elytra with the outer half and the apex testaceous and the inner margin in great part dull red. The antennae fuscous, with the first two joints light red. The underside of the head and abdomen glossy black, the thorax brownish black, and the legs light red.

The *head*, without a distinct neck and not constricted behind, is oblong, moderately convex and mainly smooth; on the forehead there are two foveae, a deeply impressed semi-circular line passes from the middle of each supra-orbital ridge to each fovea, and between the foveae is the well-marked fronto-clypeal suture. Each supra-orbital ridge bears one long posterior seta. The transverse clypeus is concave in front, and is there marked by a transverse depression and four short furrows. The labrum is narrowed, its length is nearly equal to its breadth, its straight front edge bears six long setae and there are a few short setae on each side. The antennae, which do not reach beyond the base of the thorax, are filiform; the first two segments are light red and smooth, the succeeding segments are fuscous, pubescent and setose at the distal ends; the first segment, which bears a long seta, is the longest segment, the second is the smallest, the third is less than the first but is a little longer than the succeeding segments which are about equal in length.

The *pronotum* is wider than the head, and is roughly square, with the sides slightly and angularly rounded; the front is a little wider than the base and is slightly hollowed and a little raised; the base is slightly sinuate with a very slight and wide median lobe; the disc, which is brownish black with a broad reddish testaceous border, is moderately convex and is marked by three shallow depressions, one frontal and median and one near each hind angle where the surface is also punctate. The pronotum is marginate on the sides and right round the angles; the front angles are almost sharp, the hind angles are rounded; on each side margin and near the shoulder there is a pore bearing a long seta; the faintly impressed median line does not reach the front and hind margins, and the surface of the disc is very faintly wrinkled about the median line. The *scutellum* is wanting, the specimen having been pinned through that part.

The wings are well developed. The *elytra*, which are a little broader than the thorax, are broadest at the shoulders, and are plano-convex and marginate, the bases are sinuate, the shoulders are rounded, and the apices are rounded and slightly sinuate. The outer half and the apical portion of the elytron (one is wanting) are testaceous, the inner half is brownish black, and the inner margin is in great part dull red. The surface is strongly striate, the striae converge at the apex and there unite. The scutellary striole unites with stria 1. The interstices are slightly convex and smooth. On the outermost stria and towards the apex there are three punctures and a large pore. The epipleura are broad at the shoulder.

The *legs* are light red, with the tibiae and the tarsi darker and reddish. The front and middle coxae are round; the hind coxae are flat and are not separated, and each bears a large pore. The hindmost trochanters are very long and ovoid. The front tibiae have a deep semicircular notch on the inner edge near the distal end; behind the notch is a long spine, and in front of the notch there is a much stouter terminal spine. The middle and hind tibiae have each two terminal spines. The claws are simple.

Length 7½ mm.; breadth (across the elytra) 3 mm.
Hab. America (Fab.).

Family DYTISCIDAE

The following are the species of Dytiscidae mentioned by Fabricius, in his published works, as having been described by him from specimens in Dr Hunter's Collection:

Dytiscus fasciatus *Syst. Ent.* Appendix, p. 825, Nos. 6–7 (1775).
  vittatus *Ibid.* Nos. 8–9.

The above names are the original names as given by Fabricius, and the references are to the works in which these species were first described.

17. *Hydaticus vittatus* (Fab.)

Coleopterorum Catalogus, pars 71 (A. Zimmermann, 1920), Dytiscidae, etc., p. 228.

East and South Asia, Malayan Islands, Australia.

SYN. *Dytiscus vittatus* Fab., *Syst. Ent.* Appendix, p. 825, Nos. 8–9 (1775); *Sp. Ins.* I, p. 293, No. 10 (1781); *Mant. Ins.* I, p. 190, No. 12 (1787); *Ent. Syst.* I, 1, p. 190, No. 14 (1792); *Syst. Eleuth.* I, p. 262, No. 20 (1801); Oliv. *Ent.* III, 40, p. 20, pl. 1, fig. 5 (1795).

This species is represented by two specimens under label

'*Dyt. vittatus*
Fabr. pag. 293, No. 10'

in Cabinet B, drawer 10; one is imperfect, without a head. The entire specimen is a female, and *provisionally* it may be taken as the type; it corresponds with the descriptions given by Fabricius and Olivier and with Olivier's figure, and it closely resembles the British Museum examples of *vittatus* Fab. with which it has been compared by Mr Arrow. I have compared it also with two modern examples (from Lombok) in the 'Bishop' Collection; these have on each of the abdominal sterna (second to sixth) two deep punctures or pores (one at each side), which are only slightly impressed on the type specimen.

Description of Co-type, *Dytiscus vittatus* Fab. The specimen is a *female*. Form elongate oval. Glossy black above, with a broad red band on the frons and on the lateral parts of the pronotum; and with a red band along the side of each elytron, from the shoulder to near the apex, this band being widest in the middle and double about the shoulder region, there forming a long loop incomplete in front. The underside of the body and the hindmost legs are glossy black; the sides of the prosternum, the underparts of the shoulder, the antennae, the mouth-parts, and the front legs are red; the middle legs have the femora red and the tibiae and tarsi dark reddish brown. The *scutellum* is very small and broadly triangular. On each *elytron* there is, midway between the suture and the red band, a very distinct longitudinal row of irregularly placed fine punctures; and on the inner side of the red band there is an imperfect row of fine punctures. The metathoracic episterna reach the middle coxal cavities. The sutures between the metathoracic episterna and the metasternum are straight. The spurs of the hindmost tibiae are sharp-pointed.

Length 14 mm.; breadth 7 mm.
Hab. India (Fab.), East Indies (Oliv.).

18. *Sandracottus fasciatus* (Fab.)

Coleopterorum Catalogus, pars 71 (A. Zimmermann, 1920), Dytiscidae, etc., p. 234.

Java.

SYN. *Dytiscus fasciatus* Fab., *Syst. Ent.* Appendix, p. 825, Nos. 6–7
(1775); *Sp. Ins.* I, p. 293, No. 7 (1781); *Mant. Ins.* I, p. 190,
No. 8 (1787); *Ent. Syst.* I, I, p. 189, No. 9 (1792); *Syst.
Eleuth.* I, p. 261, No. 15 (1801); Oliv. *Ent.* III, 40, p. 18,
pl. 2, fig. 19 (1795).

There are two specimens under label

'*Dyt. fasciatus*
Fabr. pag. 293, No. 7'

in Cabinet B, drawer 10; but one of these, the lighter coloured
one, is a different species (*Eretes sticticus*). The other specimen is apparently the type; it answers the descriptions of
fasciatus given by Fabricius and Olivier and corresponds with
Olivier's figure, and it closely resembles the examples of this
species in the British Museum Collection.

Description of Type, *Dytiscus fasciatus* **Fab.** Form
oval, broadening out beyond the middle. Reddish yellow
above, with two transverse brownish black marks on the
pronotum (the posterior one band-like), with a broad arcuate
indented brownish black band across the middle of the elytra,
and posteriorly a lesser arcuate brownish black band interrupted in the form of large spots, and at each apex a rounded
brownish black spot, the inner elytral borders brownish
black along the suture, and two median longitudinal rows of
more or less linear brownish black punctures (the outer one faint
and irregular) on each elytron. The underside of the body is
dull glossy blackish brown, touched with reddish brown on
the lateral parts of the abdominal sterna; the front and the
middle legs are light red, the hindmost legs are brownish
black with red trochanters. The *head* is reddish yellow,
slightly brownish black behind; the eyes are whitish yellow,
and across the middle of each eye there is a dark mark; the
antennae and the palps are yellow. The *pronotum* is rugulose.
The *scutellum* is very small and broadly triangular. The
surface of the *elytra* is finely and irregularly punctulate. On
each elytron there are two longitudinal rows of small brownish
black and wide-apart spots, and within each spot there are

from three to five contiguous punctures. On each elytral
border there is a row of punctules, which become obsolete
towards the apex. The metathoracic episterna reach the
middle coxal cavities. The sutures between the metathoracic
episterna and the metasternum are curved. The spurs of the
hindmost tibiae are sharp-pointed.

Length 14 mm.; breadth 8 mm.
Hab. India (Fab.), East Indies (Oliv.).
See Plate 10.

Super-family LAMELLICORNIA

Family SCARABAEIDAE

The following are the species of Scarabaeidae mentioned
by Fabricius, in his published works, as having been described
by him from specimens in Dr Hunter's Collection:

Scarabaeus eurytus	*Syst. Ent.* p. 7, No. 13 (1775).
aenobarbus	*Ibid.* p. 10, No. 28.
goliatus	*Ibid.* p. 13, No. 41.
gibbosus[1]	*Ibid.* p. 28, No. 112.
miliaris	*Ibid.* Appendix, p. 817, Nos. 110–111.
hastatus	*Sp. Ins.* 1, p. 6, No. 11 (1781).
quadrispinosus	*Ibid.* p. 11, No. 36.
splendidulus	*Ibid.* p. 23, No. 100.
smaragdulus	*Ibid.* p. 34, No. 157.
enema	*Mant. Ins.* 1, p. 4, No. 12 (1787).
pactolus	*Ibid.* p. 12, No. 112.
lar	*Ibid.* p. 13, No. 124.
fricator	*Ibid.* p. 15, No. 140.
Melolontha vitis	*Syst. Ent.* p. 37, No. 26 (1775).
errans	*Ibid.* No. 27.
proboscidea	*Ibid.* Appendix, p. 818, Nos. 31–32.

[1] Collection not stated by Fabricius, but probably Hunter's.

PLATE 10

Dytiscus fasciatus Fab. × 5

Melolontha innuba	*Mant. Ins.* I, p. 22, No. 45 (1787).
Trox horridus	*Syst. Ent.* Appendix, p. 818, Nos. 2–3.
Trichius viridulus	*Syst. Ent.* Appendix, p. 820, Nos. 5–6.
Cetonia tristis	*Syst. Ent.* p. 45, No. 10 (1775).
smaragdula	*Ibid.* No. 11.
haemorrhoidalis	*Ibid.* Appendix, p. 819, Nos. 34–35.
rufipes	*Mant. Ins.* I, p. 27, No. 7 (1787).
quadripunctata	*Ibid.* No. 12.
goliata	*Ent. Syst.* I, 2, p. 124, No. 1 (1792).

The above names are the original names as given by Fabricius, and the references are to the works in which these species were first described.

The following are species described by Olivier from specimens in Dr Hunter's Collection, excepting the first mentioned which is assumed to be the Drury type acquired by Hunter:

Scarabaeus oromedon	Drury, *Illus. Nat. Hist.* I, p. 81, pl. XXXVI, fig. 5 (1770).
bellicosus	Oliv. *Ent.* I, 3, p. 103, pl. 22, fig. 32 *b* (1789).
Cetonia histrio	Oliv. *Ent.* I, 6, p. 45, pl. 10, fig. 94 (1789).
bidens	*Ibid.* p. 62, pl. 10, fig. 87.
ignita	*Ibid.* p. 69, pl. 10, fig. 96.

In the following pages the above-mentioned types are described under their *modern* names and in the order adopted in Junk and Schenkling's *Coleopterorum Catalogus* and (for the *Dynastinae*) Gemminger and Harold's *Catalogus Coleopterorum*.

Sub-family COPRINAE

19. *Gymnopleurus miliaris* (Fab).

Coleopterorum Catalogus, pars 38 (J. J. E. Gillet, 1911), Scarabaeidae, Coprinae, p. 20.

India.

SYN. *Scarabaeus miliaris* Fab., *Syst. Ent.* Appendix, p. 817, Nos.
110–111 (1775); *Sp. Ins.* I, p. 32, No. 141 (1781); *Mant. Ins.*
I, p. 17, No. 162 (1787); *Ent. Syst.* I, I, p. 63, No. 209
(1792); Oliv. *Ent.* I, 3, p. 167, pl. 18, fig. 164 (1789).
Ateuchus miliaris Fab., *Syst. Eleuth.* I, p. 56, No. 5 (1801).

I could not find (in Dr Hunter's Collection) a specimen
labelled *Sc. miliaris*; apparently the original label has been
lost or destroyed. There is, however, in Cabinet A, drawer 5,
a specimen (not labelled) which is evidently the type, it
closely agrees with the descriptions of *miliaris* given by
Fabricius and Olivier and Olivier's figure resembles it.

Description of Type, *Scarabaeus miliaris* Fab.
Form short, ovate and convex. The upper surface rugulose-
punctate and closely covered with very short recumbent grey
hairs which give a shagreened effect; but there are smooth
parts which have the form of large and raised glossy black
spots amidst the grey, which now is very indistinct with age.
The lower surface dull glossy black and more or less closely
punctate. The *clypeus* rounded and broad in front and with
six tooth-like projections, the middle two being the largest;
on the surface of the clypeus three longitudinal raised lines,
the median one faint, the outer two strongly marked and
diverging widely in front, each extending to between the
smaller tooth-like projections and having about the middle
an oval spot-like thickening. The canthus or brow-ridge
prominent and so produced that it almost surrounds the
eye, only a small oval portion of which is visible from above.

The *pronotum* large and transverse, slightly wider than the
elytra, very convex and somewhat elevated, and distinctly
marginate in front and at the sides. The front of the pronotum

PLATE II

Scarabaeus miliaris Fab. × 6

has two lateral projections between which is the deep and sinuate excavation for the head; the base is straight about the middle and curves slightly forward towards the sides, which are rounded; the front and hind angles are sharp. On the surface of the pronotum are twelve large and roughly rounded glossy black spots and a small excavation at each side, the excavations being marked by two of the black spots. The *thoracic sterna* closely punctate, except about the middle where the punctures are few and farther apart; the *prosternum* has a fringe of golden hairs on its posterior border.

The *elytra* closely punctulate and distinctly marginate along the outer borders which are deeply excavated near the shoulders and there very sinuate. On the elytra is a row of seven glossy black spots at the base and a row of four in the middle; besides these, which are oblong in shape, there are two apical spots, one on each apical convexity, very large and irregular. On each side of the suture there is a parallel longitudinal stria; several other striae on the surface have the form of long wavy loops within each other and with pointed bends.

The *abdominal sterna* punctulate and lightly fringed with fine hairs, except the middle portions which are more or less smooth. The *pygidium* is closely punctate.

The *legs* have the femora punctulate. The hind legs are long. The front tibiae have each three blunt tooth-like processes, a small articulated spine and a short tarsus with simple claws; outer edges of the front tibiae finely serrate between and behind the tooth-like processes. The middle tibiae have each three small tooth-like processes and a long terminal articulated spine. The hind tibiae have serrated outer edges and a long terminal articulated spine.

Length 10 mm.; breadth (across the pronotum) 6 mm.
Hab. India (Fab. and Oliv.).
See Plate 11.

20. *Canthon smaragdulus* (Fab).

Coleopterorum Catalogus, pars 38 (J. J. E. Gillet, 1911), Scarabaeidae, Coprinae, p. 33.

Brazil.

SYN. *Scarabaeus smaragdulus* Fab., *Sp. Ins.* I, p. 34, No. 157 (1781);
Mant. Ins. I, p. 18, No. 179 (1787); *Ent. Syst.* I, I, p. 70,
No. 234 (1792); Oliv. *Ent.* I, 3, p. 159, pl. 14, fig. 131 (1789).
Ateuchus smaragdulus Fab., *Syst. Eleuth.* I, p. 58, No. 17
(1801).

The specimen under label

'*Sc. smaragdulus*
Fabr. pag. 34, No. 127'

in Cabinet A, drawer 1, is not the one described as *smaragdulus* by Fabricius. In the same drawer, however, there is a
specimen under label

'*Sc. aenobarbus*
Fabr. pag. 10, No. 32'

which agrees with the description of *smaragdulus* given by
Fabricius; it agrees also with Olivier's description of that
species and it closely corresponds with his figure. Mr Arrow
has examined this specimen and he confirms my opinion that
it is the misplaced type. Comparing the type with modern
examples of the species in the British Museum, we find that
the species commonly known as *smaragdulus* is not *smaragdulus* but is *speculifer*. *Canthon speculifer* will now have to take
the name of *smaragdulus*; and, consequently, the supposed
smaragdulus is meanwhile without a name.

Description of Type, *Scarabaeus smaragdulus* Fab.
Form ovate and convex. The upper surface uniformly dark
greenish blue and very brilliant; the lower surface blackish
green and shining, except the metasternum and the femora
which are dark greenish blue. *Head* about half the breadth
of the thorax; the vertex a little elevated and rounded, convex;
the clypeus rounded in front and bearing two prominent and
contiguous median tooth-like projections, and the edge sin-

PLATE 12

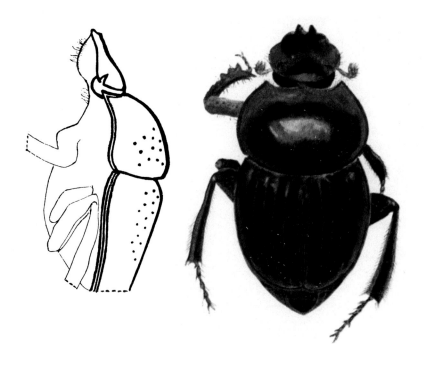

Scarabaeus smaragdulus Fab. × 5
and side view of head and pronotum

uate at each side of these. The canthus or brow-ridge prominent and so produced that it almost covers the (brown)
eye, only a small part of which is visible from above. The
clubs of the antennae are light red.

The *pronotum* very convex, somewhat elevated, its breadth
equal to that of the middle and broadest part of the elytra;
the front of the pronotum hollowed out between the sharp
front angles, the base curving forward towards the sides
which are rounded and marginate, the hind angles blunt, the
surface smooth and shining (glabrous). The *prothoracic episterna* punctate, the punctures bearing short brownish hairs.
The *metasternum* strongly convex, smooth and shining.

The *elytra* short, not covering the whole of the abdomen,
a little flattened, narrowed towards the base and also towards
the apices which are convex and blunt and slightly turned
forwards at the suture; the outer borders rounded and marginate, the shoulder callus prominent, the surface mainly
smooth and shining (glabrous) and marked with very faint
longitudinal striae which are, however, strongly impressed at
the base.

The *abdominal sterna* are smooth and glossy. The *pygidium*,
smooth and glossy, resembles a short and blunted cone.

The *legs* are moderately long; the femora and tibiae of the
front legs are distinctly granulate on the upper surface; the
front tibiae have each three blunt tooth-like processes, a small
terminal articulated spine and a short tarsus with two simple
claws; the middle and the hind tibiae have each a long terminal articulated spine; the femora of the hind legs are
punctulate.

Length 15 mm.; breadth 8 mm.
Hab. South America (Fab. and Oliv.).
See Plate 12.

21. *Deltochilum gibbosum* (Fab.)

Coleopterorum Catalogus, pars 38 (J. J. E. Gillet, 1911), Scarabaeidae, Coprinae, p. 36.

Carolina.

SYN. *Scarabaeus gibbosus* Fab., *Syst. Ent.* p. 28, No. 112 (1775);
Sp. Ins. 1, p. 32, No. 143 (1781); *Mant. Ins.* 1, p. 17,
No. 164 (1787); *Ent. Syst.* 1, 1, p. 64, No. 213 (1792);
Oliv. *Ent.* 1, 3, p. 154, pl. 16, fig. 151 *b* (1789).
Ateuchus gibbosus Fab., *Syst. Eleuth.* 1, p. 57, No. 13 (1801).

Fabricius does not say (in his descriptions of *Scarabaeus
gibbosus*) to which collection the type belongs; but this species
was also described (and figured) by Olivier from a specimen
"du Cabinet de feu M. Hunter". I think we may assume
that the Hunterian insect is the actual Fabrician type, unless
direct evidence to the contrary is forthcoming.

Two specimens under label

> '*Sc. gibbosus*
> Fabr. pag. 32, No. 143'

in Cabinet A, drawer 1, are obviously not that species as
described by Fabricius and Olivier. In the same drawer
there are two specimens under label

> '*Sc. goliatus*
> Fabr. pag. 14, No. 51'

neither of these is *goliatus*, but one of them undoubtedly is
Scarabaeus gibbosus and, in all probability, it is the example
described and figured by Olivier.

Description of Type, *Scarabaeus gibbosus* Fab.
Form short and broad, gibbous; uniformly dull brownish
black. *Head* transverse, half the breadth of the thorax, transversely ovate, flattened and punctate; the *clypeus* reflexed in
front and there emarginate with four tooth-like projections,
two at each side of a wide median hollow, the inner ones more
pointed than the outer ones; posteriorly the clypeus is distinctly marked off from the vertex by a strong transverse

PLATE 13

Scarabaeus gibbosus Fab. $\times 3\frac{1}{2}$
and head (dorsal aspect)

ridge which extends between the dark brown eyes. The canthus or brow-ridge produced horizontally in front of each eye and extending backwards as a short process almost touching the pronotum, thus dividing the eye into upper and lower portions. Clubs of the antennae light red.

The *pronotum* large and broad, almost as broad as the widest part of the elytra, and distinctly marginate all round; the front with two lateral projections and, between these, a broad excavation for the head; the sides angular, the base rounded and slightly sinuous; very convex above, the surface punctate, with ringed (annular) punctures except about the centre; on the surface three small tubercles, one at each side (opposite the angle) and one in the middle of the margin of the base. The *thoracic sterna* punctate, except about the middle of the metasternum which is smooth and glossy.

The *elytra* short, not covering the abdomen entirely, narrowed at the base, broadest about the middle and abruptly rounded at the apex; the inner and outer borders slightly marginate, the outer borders rounded and sinuate; on each side of the suture and near the base a prominent, rounded, sub-conical elevation which renders the form characteristically gibbous; at the base on each side three strongly marked longitudinal ridges, the inner one short and the outer one extending almost half the length of the elytra, and upon each apex five very short and oblique ridges which together form prominent apical elevations; the surface punctate, rugulose, with longitudinal rows of small tubercles, and costate, the costae rather faint except at the base and the apices.

The *abdominal sterna* more or less punctate; the sternum of the first segment of the abdomen has a very prominent enlarged keel which underlies and conceals the next three sterna. The *pygidium* punctate.

The front *femora* are broad and flattened, and at the base of each there is (according to Fabricius and Olivier) a red or brown spot. The front *tibiae* have each three blunt tooth-like projections on the outer side and two on the inner side, and

a blunt terminal articulated spine; the edges of the front
tibiae are serrated. The middle tibiae are square-sided, and
each has two terminal spines. The hind *legs* are very long,
longer than the other two pairs, and are bent (incurved);
the long and incurved hind tibiae have prominent longitudinal ridges and each has a terminal articulated spine.

Length 23 mm.; breadth 18 mm.
Hab. America (Fab. and Oliv.).
See Plate 13.

22. *Copris fricator* (Fab).

Coleopterorum Catalogus, pars 38 (J. J. E. Gillet, 1911), Scarabaeidae, Coprinae, p. 74.

British East Indies.

SYN. *Scarabaeus fricator* Fab., *Mant. Ins.* 1, p. 15, No. 140 (1787);
Ent. Syst. 1, 1, p. 54, No. 176 (1792); Oliv. *Ent.* 1, 3, p. 122,
pl. 16, fig. 149 (1789).
Copris fricator Fab., *Syst. Eleuth.* 1, p. 45, No. 67 (1801).
C. indicus Gillet.

Two specimens under label

'*Sc. fricator*
Fabr. MSS'

in Cabinet A, drawer 1, are obviously not that species as
described by Fabricius. In the same drawer there are two
beetles which have been misplaced under label *Sc. ammon*;
one of these is a male of *Bubas bison* and the other is evidently
the missing type of *fricator* Fab. It closely agrees with the
descriptions of *fricator* given by Fabricius and Olivier, and it
corresponds with Olivier's figure, which is, however, a poor
representation.

I have compared this insect with an example of *Copris
sinicus* (*sulcicollis*) from the British Museum Collection, because Mr Arrow thinks that *sinicus* may prove to be *fricator*
and that modern examples at present known as *fricator* are not
that species. The following are the features of difference
between this type and *sinicus*: The median longitudinal line

on the thorax is only slightly impressed, whereas in *sinicus* it is deeper and broader, a very distinct furrow. The lines of transverse double dots on the elytra are more prominent than those of *sinicus* and have a crenulate-striate appearance. The transverse elevation of the thorax is interrupted in the middle and has not so distinct a ridge as in *sinicus*, and at each side of the ridge there is a small round tubercle which is not present in *sinicus*; the lateral depressions are oval, not round as in *sinicus*, and the projecting line between the lateral depression and the lateral margin is prominent, not slight as in *sinicus*.

Description of Type, *Scarabaeus fricator* **Fab.** Form oblong, stout, sub-cylindrical, convex; uniformly dull black. The *head* transverse, nearly as broad as the thorax, flattened, punctate and a little rugulose, marginate and reflexed in front; the clypeus broad, rounded and reflexed in front and there slightly sinuate and emarginate (a cleft in the middle); the clypeus is distinctly marked off from the vertex by a transverse and semicircular suture with a slight ridge, and the middle part of this ridge is elevated into a short and thick erect horn, rounded in front and flattened behind, truncate and slightly notched. The canthus or brow-ridge rugulose-punctate, broadly produced and almost completely surrounding the eye, only an oval portion of which is visible from above.

The *pronotum* large and transverse, its breadth almost equal to that of the elytra, marginate all round and with the greater part of the front margin broadly flattened out; elevated in front (almost vertical) and very convex, the surface closely punctate and with rugulose effect; the front with a broad and deep excavation for the head; the sides a little sinuate, with the front angles sharp and the hind angles rounded; the base rounded, curving gently forward to the sides; the transverse frontal elevation forms a blunt median ridge, interrupted in the middle and with a small round tubercle near each end; in the middle of the convex disc there is a short and slightly impressed median longitudinal line, and on each side there is an oval depression beside which a prominent projecting ridge line extends obliquely outward and forward to the lateral margin. The *thoracic sterna* are punctate.

The *elytra* are a little broader than the pronotum, very convex, slightly narrowed at the base, gradually rounded at the apices, and marginate; the outer borders are inflexed, sinuate and strongly marginate; the shoulder callus is distinct, the apical callus is prominent; the surface is faintly punctulate and is distinctly costate, and between the costae there are longitudinal furrows (crenulate-striate) with rows of transverse double punctures; the stria on each side of the suture extends to the apex, the other striae (with the exception of two which become obsolete at the apical callus) have the form of long loops within each other and with sharply pointed apical bends.

The *abdominal sterna* and the *pygidium* are punctate.

The *legs* are short; the front femora are broad and rounded, and on the base of each there is a large puncture. The front tibiae have each three blunt tooth-like projections on the outer side and a terminal blade-like articulated spine. The middle tibiae, which are greatly expanded at the distal ends, have serrate edges, and each has two terminal articulated spines, one longer than the other. The hind tibiae have the outer part of the expanded distal end greatly excavated, and each has one terminal articulated spine.

Length 18 mm.; breadth (across the elytra) 10 mm.
Hab. Eastern India (Fab.), East Indies (Oliv.).
See Plate 14.

23. *Copris minutus* (Drury)

Coleopterorum Catalogus, pars 38 (J. J. E. Gillet, 1911), Scarabaeidae, Coprinae, p. 76.

North America.

SYN. *Scarabaeus minutus* Drury, *Illus. Exot. Ins.* 1, p. 78, pl. XXXV, fig. 6 (1770).
    S. silenus Fab., *Syst. Ent.* p. 21, No. 83 (1775).

PLATE 14

Scarabaeus fricator Fab. × 4

S. lar Fab., *Mant. Ins.* I, p. 13, No. 124 (1787).
S. ammon Fab., *Sp. Ins.* I, p. 24, No. 105 (1781); *Ent. Syst.*
    I, I, p. 44, No. 147 (1792); Oliv. *Ent.* I, 3, p. 124, pl. 12,
    fig. III (1789).
Copris ammon Fab., *Syst. Eleuth.* I, p. 35, No. 25 (1801).

This is a pre-Fabrician type, the *male* having been first
described by Drury under the name of *Scarabaeus minutus*.
It was later described by Fabricius three times from examples
in different collections and under different names. Olivier
pointed out that the insect described by Fabricius as *lar*,
from specimens in Dr Hunter's Cabinet, was the same as the
one which Fabricius described later as *ammon*, from specimens in the Banks Collection. In his *Entomologia Systematica* Fabricius recognised the synonymy of his *silenus*, *lar*
and *ammon*; but he did not mention *minutus* Drury.

Two specimens under label

'*Sc. lar*
Fabr. MSS'

in Cabinet A, drawer I, are evidently co-types of the male of
lar Fabricius, they answer the description of that species given
by Fabricius; they correspond also with the description and
figure of *ammon* given by Olivier, and with the description
and figure of *minutus* given by Drury.

Description of Co-type, *Scarabaeus lar* Fab. *Male.*

Form elongate, sub-cylindrical, very convex; uniformly dull
black. The *clypeus* smooth; the front broad, rounded, reflexed and emarginate (a cleft in the middle); posteriorly a
short, stout, simple and erect frontal horn, very slightly recurved at the tip and projecting slightly forwards. The canthus
or brow-ridge punctate, prominent and extending broadly
and horizontally in front of each eye, and produced backwards as a short process almost reaching the pronotum, thus
dividing the (light brown) eye into an oval upper portion and
a large lower portion. The clubs of the antennae ferruginous.

The *pronotum* large and transverse, its breadth equal to
that of the elytra, marginate and punctate; elevated from

behind forwards, abruptly perpendicular in front and with three distinct frontal prominences, a broad median one between two which are small and tubercle-like; on each side a small depression beside a slight and obliquely directed ridge. The front of the pronotum with two broad lateral projections and, between these, a slight excavation for the head; the sides gently rounded, the front and hind angles rounded off; the base a little rounded, curving gently forward towards each side; immediately in front of the base a transverse furrow, from the middle of which a median longitudinal impressed line extends to the median frontal prominence. The *thoracic sterna* punctate, except the middle portion of the metasternum which is smooth, and upon which there is a short and deeply impressed longitudinal line.

The *elytra* a little narrowed at the base, gradually rounded at the apices, the outer borders inflexed and slightly marginate; the surface striate-punctate, marked with deeply impressed longitudinal furrows, each with a double row of punctures and having a crenulate-striate appearance; the stria on each side of the suture extends to the apex, the other striae have the form of long loops within each other and with sharply pointed apical bends.

The *abdominal sterna* and the *pygidium* are closely punctate.

The *legs* are short; the front tibiae have each four tooth-like processes (the proximal one very small), a terminal articulated spine, and a small tarsus with simple claws; the middle and the hind tibiae are greatly expanded at the distal ends and each has two long terminal articulated spines.

Length 12 mm.; breadth 6 mm.
Hab. Eastern India, America (Fab.), New York (Drury).
See Plate 15.

24. *Circellium bacchus* (Fab.)

Catalogus Coleopterorum (Gemminger and Harold, 1869), IV, Scarabaeidae, p. 983.

South Africa.

PLATE 15

Scarabaeus lar Fab. ♂ × 5

SYN. *Scarabaeus bacchus* Fab., *Sp. Ins.* I, p. 32, No. 142 (1781);
    Mant. Ins. I, p. 17, No. 163 (1787); *Ent. Syst.* I, 1, p. 64,
    No. 212 (1792); Oliv. *Ent.* I, 3, p. 153, pl. 17, fig. 161 (1789).
    Ateuchus bacchus Fab., *Syst. Eleuth.* I, p. 57, No. 12 (1801).
    A. hemisphaericus Pallas, *Icon.* p. 20, t. B. f. 23.

The type of this species is stated by Fabricius to be in the
Schulz Collection. Olivier described and figured a specimen
(from Cape of Good Hope) in Dr Hunter's Collection. This
specimen is noted in the card-index as missing. There is,
however, in Cabinet C, drawer 32, an insect (not labelled)
which corresponds perfectly with the descriptions of *bacchus*
given by Fabricius and Olivier, and Olivier's figure closely
resembles it. I have also compared it with modern examples
of this species in the British Museum Collection; and Mr
Arrow confirms its identity.

25. *Phanaeus bellicosus* (Oliv.)

Catalogus Coleopterorum (Gemminger and Harold, 1869), IV,
    Scarabaeidae, p. 1017.

Brazil.

SYN. *Scarabaeus bellicosus* Oliv., *Ent.* I, 3, p. 103, pl. 22, fig. 32 *b*
    (1789).
    S. silvanus Dej.
    S. sylvanus Casteln.

The type of this species is noted in the card-index of Dr
Hunter's Collection as missing; but in Cabinet E, drawer 32,
there is a specimen (not labelled) which answers the description given by Olivier and which resembles his figure. It also
closely matches the modern examples of this species in the
British Museum Collection, with which I have compared it,
and evidently it is the Olivier type.

Description of Type, *Scarabaeus bellicosus* Oliv.
Form short and stout, sub-cylindrical. Glossy black, with
touches of violet on the head and the pronotum. The *head* is
half the breadth of the pronotum. The clypeus is rugulose
and is rounded and strongly marginate in front, with the
margin upturned and with two conspicuous pointed inden-

tations; posteriorly a long erect and simple horn (length half an inch), recurved towards the pointed tip, slightly toothed behind, cylindrical except about the base which is wide and thick and frontally flattened. A prominent ridge extends obliquely outwards from the horn to the lateral margin and marks off the clypeus from the canthus, which is broadly produced around the front part of the eye; the surface of the canthus is rugose-punctate and in great part bright violet coloured.

The *pronotum* is a little wider than the elytra, transverse, strongly marginate all round, widely excavated about the middle of the disc and elevated behind; in front there is a shallow excavation for the head; the base is angular, the rounded sides have a large and deep rounded notch posteriorly, the front angles are rounded and the hind angles are sharp. The pronotal elevation is horseshoe-shaped, with an oval depression at each end and a deep cavity in the middle between two raised ridges, which resemble short and compressed horns, each with two blunt tips and united together in front. The surface of the pronotum is mainly rugulose and is marked with violet.

The *elytra* are very convex and marginate, broadest at the base, narrowed towards the apex and rectangular at the posterior angles; the surface is marked with parallel striae between the costae, which are imperfect and slight except at the base; a raised and thick ridge-like line extends along each side of the suture from the apex to about the middle; the apical callus is very prominent, the shoulder callus less so.

The *thoracic sterna* are more or less punctulate and hairy with reddish coloured hairs. The *pygidium* is punctulate and bright violet coloured.

The front *tibiae* have each four blunt tooth-like projections and a terminal hook-like articulated spine; the middle and hind tibiae have the distal ends broadly expanded and have each a terminal three-sided blade-like articulated spine.

Length 34 mm.; breadth (across the pronotum) 22 mm.
Hab. Cayenne (Oliv.).
See Plate 16.

PLATE 16

Scarabaeus bellicosus Oliv. × 2½

26. *Phanaeus splendidulus* (Fab.)

Coleopterorum Catalogus, pars 38 (J. J. E. Gillet, 1911), Scarabaeidae, Coprinae, p. 86.

South Brazil, Uruguay, Argentina.

SYN. *Scarabaeus splendidulus* Fab., *Sp. Ins.* 1, p. 23, No. 100 (1781); *Mant. Ins.* 1, p. 12, No. 110 (1787); *Ent. Syst.* 1, 1, p. 42, No. 138 (1792); Oliv. *Ent.* 1, 3, p. 111, pl. 2, figs. 18 *a* and *b* (1789).
Copris splendidulus Fab., *Syst. Eleuth.* 1, p. 32, No. 8 (1801).

There are two specimens under label

'*Sc. splendidulus*
Fabr. pag. 23, No. 100'

in Cabinet A, drawer 1; one is a male, the other is a female. The male is the type; it agrees closely with the description given by Fabricius, and it also agrees with the description of the male given by Olivier. Olivier described and figured both sexes from specimens (said to be from Madagascar) in the Royal Cabinet at Paris; his figure of the female resembles it, but is badly drawn.

The modern examples of *splendidulus* in the British Museum are not like the Fabricius type and are therefore not that species; the modern species which agrees with the type is *Phanaeus floriger* Kirby, of Southern Brazil, the male of which has the horn of the head *truncate* at tip, not pointed, and the pronotal elevations *produced into horn-like processes*. *Phanaeus floriger* will now have to take the name of *splendidulus*; and, consequently, the species commonly known as *splendidulus* is meanwhile without a name.

Description of Type, *Scarabaeus splendidulus* Fab. *Male.* Form short and stout, sub-cylindrical. *Head* about half the breadth of the pronotum. The *clypeus* rugulose and dark bronze-green with an interrupted band of bright coppery green at the base of the horn, which is dark bluish green; the front rounded and marginate, a slight notch in the middle and a fringe of short hairs beneath the front margin;

posteriorly a long and erect horn, recurved towards the tip, with the base wide and thick and frontally flattened and bearing two small and curved frontal elevations, together resembling an inverted V, with the tip frontally flattened out, truncate, a little broader than the middle portion, and slightly cleft. The canthus or brow-ridge broadly produced and almost surrounding the eye, only an oval portion of which is visible from above.

The *pronotum* smooth and bright coppery green, large and convex, its breadth greater than its length, and strongly marginate all round; the front with two broad lateral lobe-like projections between which is the excavation for the head; the base angular, the sides very sinuate (sickle-shaped), the front angles rounded and the hind angles sharp. Two postero-lateral sharp pyramidal elevations of the pronotum are produced into high horn-like processes with broad and thin tips, which are incurved and somewhat intwisted; and between these horns there are two deep and rounded hollows, confluent but distinctly defined behind by an intermediate small and flat diamond-shaped area situated at the angle in the middle of the base of the pronotum. At the front of the pronotum a small median ridge-like elevation; and on each side, about the middle and near the margin, a small excavation.

The *elytra* dull bronze-green, slightly narrowed at the base and gradually rounded at the apices, the inner and outer borders marginate and the outer margins slightly inflexed; the surface strongly costate, the sutural costa obsolete towards the base and narrower but more prominent than the others, and bright coppery green, with deeply impressed and parallel longitudinal lines or striae between the costae; the stria along each side of the suture extends to the apex, the other striae become obsolete before the apex.

The *underparts of the body* glossy bronze-green with a bluish iridescence, except the femora which are bright coppery green. The *thoracic sterna* punctate and hairy, except the middle portion of the metasternum which is faintly

PLATE 17

Scarabaeus splendidulus Fab. ♂ × 4

punctulate (almost glabrous), flat and roughly diamond-shaped. The *abdominal sterna* punctate, but the middle portions punctulate with few and faintly impressed punctures. The *pygidium* bright coppery green and faintly punctulate (almost glabrous).

The *legs* moderately long and fringed with dark hair; the femora spindle-shaped and bright coppery green; the front tibiae have each three blunt tooth-like projections and a terminal blade-like articulated spine; the middle and the hind tibiae have the distal ends broadly expanded, the middle tibiae have each two terminal articulated spines and the hind tibiae have one.

Length 21 mm.; breadth 13 mm.
Hab. South America (Fab. and Oliv.).
See Plate 17.

The *female* is smaller than the male and not so broad, and is darker in colour, the head being a dull coppery green, the thorax and elytra bluish green. The *head* is like that of the male, but has a strongly raised transverse arcuate line instead of a horn. The shape of the thorax is the same as in the male; at the front of the *pronotum* there is a prominent raised triangular ridge, immediately behind which and occupying the angle is a deep and round cavity or fossa; on each side, about the middle and near the margin, there is (as in the male) a small excavation, and posteriorly there is a slight and broad median furrow.

Length 19 mm.; breadth 11 mm.

27. *Onthophagus pactolus* (Fab.)

Catalogus Coleopterorum (Gemminger and Harold, 1869), IV, Scarabaeidae, p. 1034.

Madras.

SYN. *Scarabaeus pactolus* Fab., *Mant. Ins.* I, p. 12, No. 112 (1787); *Ent. Syst.* I, 1, p. 42, No. 140 (1792); Oliv. *Ent.* I, 3, p. 119, pl. 16, figs. 144 *a* and *b* (1789).
Copris pactolus Fab., *Syst. Eleuth.* I, p. 33, No. 12 (1801).

There are two specimens under label

'*Sc. pactolus*
Fabr. MSS'

in Cabinet A, drawer 1, and these, I find, answer exactly
to the descriptions of this species given by Fabricius and
Olivier. One of the specimens is a male; it is the one described by Fabricius and is therefore the type. The other
specimen is a female of the same species; the female of
pactolus was described and figured by Olivier from a specimen
in the *Lee* Collection. Olivier's figure of the type male is
a fairly accurate representation; and his figure of the Lee
female resembles the Hunterian example.

Description of Type, *Scarabaeus pactolus* **Fab.**
Male. Form ovate and convex; brassy above, bright bronze-
green beneath, the underparts and the legs fringed with
yellowish hairs. *Antennae* light coral red. The *head* armed
with a long and erect horn, recurved towards the blunt-
pointed tip, and with two small tooth-like processes which
arise from the sides of the horn, a little beyond the middle
of its length, and which project backwards and outwards;
the lower half of the horn flattened in front and on the sides
and convex behind, the upper half rounded (cylindrical).
The *clypeus* broad and rounded in front, punctulate with hairs
set in the punctures, and covered with a dense pubescence
of short and recumbent light golden hairs. The canthus or
brow-ridge prominent, forming a lateral continuation of the
clypeus, and so produced that it completely surrounds the
eye, only a small part of which is visible above.

The *pronotum* conspicuously large, its breadth in front
almost equal to its breadth behind, distinctly marginate, and
punctulate with hairs set in the punctures; the front with a
broad and deep excavation for the head between two lateral
lobe-like projections; the sides rounded and the base roughly
angular, the middle of the base forming a distinct lobe; the
front angles blunt and the hind angles rounded off. Two

PLATE 18

Scarabaeus pactolus Fab. ♂ × 5½

elevations, culminating in two small dark green ridge-points, together form the prominent hump representing the posterior and greater part of the pronotum; between the elevations a wide longitudinal furrow continuous with the depression of the anterior part of the pronotum and marked by a slightly impressed median line; in front two small depressions beside the bases of the frontal lobes. The *thoracic sterna* punctate and more or less clothed with short fulvous hairs.

The *elytra* dull reddish brown, finely punctate and striate, and lightly clothed with a fine pubescence of very short hairs (like that of the pronotum); along the suture the inner edges of the elytra form a raised line.

The *abdominal sterna* punctulate and with a few hairs, except on the middle parts which are smooth and glossy. The *pygidium* punctulate and with a delicate pubescence.

The front *tibiae* quadridentate, having each four blunt tooth-like projections, the proximal one very small; the middle and hind tibiae have each a long articulated spine at the distal end.

Length 15 mm.; breadth (across the pronotum) 9 mm.
Hab. Brazil (Fab. and Oliv.), Madras (Gemm. and Har.).
See Plate 18.

The *female* has the *head* armed with three very small horns which arise close together as mere point-like projections of a slightly raised transverse line, the middle one a little larger than the other two.

The front part of the *pronotum* is greatly reduced, a very small and depressed area, with a slight and sinuous ridge-elevation representing two confluent ridge-points on its posterior border; the greater part of the pronotum is elevated and convex, with a very faintly impressed median longitudinal line ending posteriorly in a slight and short furrow, and with a broad and shallow depression along each side.

The front *tibiae* have each four blunt tooth-like projections (like those of the male and a terminal articulated spine which is outcurved and blade-like; and there is a small tarsus with two simple claws.

Length 12 mm.; breadth (across the pronotum) 8 mm.

Sub-family TROGINAE

28. *Trox* (*Phoberus*) *horridus* Fab.

Coleopterorum Catalogus, pars 43 (Gilbert J. Arrow, 1912), Scarabaeidae, Troginae, etc., p. 57.

Cape Colony.

SYN. *Trox horridus* Fab., *Syst. Ent.* Appendix, p. 818, Nos. 2–3 (1775); *Sp. Ins.* I, p. 34, No. 3 (1781); *Ent. Syst.* I, I, p. 87, No. 5 (1792); *Syst. Eleuth.* I, p. 111, No. 7 (1801); Oliv. *Ent.* I, 4, p. 5, pl. I, fig. 2 (1789).
Scarabaeus pectinatus Pall.
S. silphoides Thunb.

The specimens (three in number) under label

'*Trox horridus*
Fabr. pag. 34, No. 3'

in Cabinet A, drawer 2, are not *horridus*. In the same drawer, however, there are two insects under label

'*Cet. hirta*'

which are certainly not that species but which are unquestionably the misplaced type specimens of *horridus* Fab. I have examined and compared them with the descriptions of this species by Fabricius and Olivier, and with the recent and more detailed description by Péringuey (*Trans. S. Afr. Phil. Soc.* XII, 1901, p. 460). They have also been compared with modern examples in the British Museum, and their identity has been confirmed by Mr Arrow.

The locality mentioned by Fabricius is India; Olivier, in his description of a specimen in the Banks Collection, said East Indies.

Description of Co-type, *Trox horridus* Fab. Form oblong-ovate, ridged above and bristly; dull black. The *head* very rugose, especially towards the vertex, and with two transverse ridges, the upper one interrupted but more prominent, having the form of two transverse tubercles each set with a row of short and stiff bristles; the clypeus somewhat blunt at the tip; the clubs of the antennae reddish.

The *pronotum* very convex, transverse, with rounded sides which are widely marginate, the margins being produced as flat plates with arcuate edges fringed with closely set short and stiff bristles. The surface of the pronotum very rugose and deeply furrowed; the ridges between the median furrow and the curved lateral furrows thick and rounded and set with short stiff cylindrical bristles and bearing a small tubercle on their outer sides anteriorly; at each side between the ridge and the margin there is a large elongate tubercle which extends forward from the base of the pronotum to near the sinuate front. The *scutellum* is small and triangular.

The *elytra*, which cover the abdomen entirely, are narrowed at the base and sloping at the shoulders, widening out about the middle and from there rounded in to the apices; the outer margins are serrate, set with a regular series of tubercles each bearing a fascicle of short bristles; the surface of each elytron is serrate-costate, having five longitudinal rows of fasciculate tubercles, and the intervals between the rows are marked by transverse foveae which are divided by a more or less distinct line of small tubercles.

The *meso-* and *meta-sterna* have triangular projections between the intermediate coxae. The *coxae* are contiguous, and those of the middle legs are sub-globose. The front *tibiae* have each four tooth-like projections and a terminal articulated spine; there is a slight sinuation between the two distal tooth-like projections, and these are separated from the proximal two by a wide and well-marked sinuation; before the first tooth-like projection the outer edge of each front tibia is slightly serrate. The middle and hind tibiae have each five ridges set with bristles.

Length 18 mm.; width (across the elytra) 11 mm.
Hab. India (Fab.), East Indies (Oliv.).

Sub-family MELOLONTHINAE

29. *Peritrichia proboscidea* (Fab).

Coleopterorum Catalogus, pars 50 (K. W. von Dalla Torre, 1913), Scarabaeidae, Melolonthinae IV, p. 342.

Cape Colony.

SYN. *Melolontha proboscidea* Fab., *Syst. Ent.* Appendix, p. 818, Nos. 31–32 (1775); *Sp. Ins.* I, p. 44, No. 49 (1781); *Ent. Syst.* I, 2, p. 175, No. 84 (1792); *Syst. Eleuth.* II, p. 179, No. 111 (1801); Oliv. *Ent.* I, 5, p. 59, pl. 8, fig. 96 (1789).

The type of this species is noted in the card-index of Dr Hunter's Collection as missing. I have failed to trace it.

Sub-family RUTELINAE

30. *Pelidnota chamaeleon* var *ignita* Oliv.

Coleopterorum Catalogus, pars 66 (F. Ohaus, 1918), Scarabaeidae, Rutelinae, etc., p. 29.

Cayenne, Brazil.

SYN. *Cetonia ignita* Oliv., *Ent.* I, 6, p. 69, pl. 10, fig. 96 (1789).

The specimen under label

'*Cetonia ignita* Oliv.'

(the handwriting presumably Olivier's) in Cabinet A, drawer 2, is obviously not *ignita* Oliv. At the top of the same drawer and under label

'*Cet. chinensis*'

there is a specimen which is not that species, but which answers the description of *ignita* given by Olivier, and it also resembles his figure of that insect. Mr Arrow and I have compared and identified this specimen with modern examples in the British Museum Collection; it is evidently the type, either the label or the insect having been misplaced.

Description of Type, *Cetonia ignita* Oliv. Form obovate; plano-convex above. The upper surface of the body brilliant coppery gold; the under surface similarly coloured, closely punctate and clothed with fringes of light reddish hair, except the thoracic sterna which are feebly punctulate and smooth. The legs bright coppery gold, except the tarsi which are dark bluish green. The antennae and the maxillae dark bluish green; the large rounded gula light red.

The *head* is moderately large and is slightly convex; the clypeus is broadly triangular, its front edge is upturned, its apical margin is very feebly cleft, and the eye-ridges are pointed. There is a double shallow depression in the median lobe of the sinuate fronto-clypeal carina, which is clearly marked. The surface of the clypeus is strongly rugose in front and is closely punctate, except towards the frons where the punctures are fewer and scattered. The maxillae have each two prominent, broad and sharp-edged teeth.

The *pronotum* is convex; the surface is punctulate and has two conspicuous mid-lateral depressions marked with large punctures close together. The front of the pronotum is narrowed and excavate and has a distinct median lobe-like projection, the sides are very slightly angular, and the base is sinuate with a slight and broad median lobe; the front angles are acute, the hind angles are rounded. The pronotum is strongly marginate, except at the frontal lobe where the border is interrupted.

The *scutellum* is heart-shaped and smooth.

The *elytra*, narrowed a little towards the base and gently rounded at the apices, do not cover the whole of the abdomen, and are marginate all round, the outer border slightly upturned. The surface is distinctly but not strongly costate with longitudinal impressed lines of punctures bordering the costae and continuous as loops at the prominent apical callus; in the interval between the sutural and the second costae there is an irregular double row of punctures which become obsolete towards the apex; the lateral costae and intervals are more or less obsolete between the prominent shoulder callus and the outer margin where the surface is very corrugated.

The prominent *pygidium* is closely punctate and thickly covered with light reddish hair. The front and hind *coxae* are contiguous. The *prosternum* is produced downwards behind the front coxae and is bent forward at a right angle as a flattened trowel-shaped process, which partly conceals the junction of the coxae. The *mesosternum* is produced as a blunt-pointed process between the middle coxae and is marked by a deep median longitudinal line. The hindmost *femora* are very broad. The front *tibiae* have each three tooth-like projections and an articulated spine. The middle and the hind tibiae are somewhat triangular in shape, widest at the distal ends; and the outer surfaces are convex, the inner surfaces flattened. On the outer surfaces of the middle and hind tibiae there are two oblique rows of small spines; and the distal ends have a fringe of short spines and (at the inner edge) two larger spines. The middle *tarsi* have two long and simple curved claws, the outer one stouter and longer than the inner one and with a slight indication of a little tooth on its under edge. The claws of the other tarsi are wanting in this type specimen.

Length 36 mm.; breadth (across the elytra) 18 mm.
Hab. Sonth America, Cayenne and Surinam (Oliv.).
See Plate 19.

31. *Antichira virens* (Drury)

Coleopterorum Catalogus, pars 66 (F. Ohaus, 1918), Scarabaeidae, Rutelinae, etc., p. 51.

Cayenne, Surinam, Obidos.

Syn. *Scarabaeus virens* Drury, *Illus. Exot. Ins.* ii, p. 54, pl. XXX,
fig. 3 (1773).
Cetonia smaragdula Fab., *Syst. Ent.* p. 45, No. 11 (1775);
Sp. Ins. i, p. 53, No. 14 (1781); *Mant. Ins.* i, p. 28, No. 22
(1787); *Ent. Syst.* i, 2, p. 134, No. 34 (1792); *Syst. Eleuth.*
ii, p. 143, No. 44 (1801); Oliv. *Ent.* i, 6, p. 73, pl. 10,
fig. 90 (1789).
C. smaragdina Linn.

Another instance of a misplaced type in the Hunterian Collection. The insect under label

'*Cet. smaragdula*
Fabr. pag. 53, No. 14'

PLATE 19

Cetonia ignita Oliv. × $2\frac{1}{4}$

in Cabinet A, drawer 2, is not the one described as such by Fabricius and Olivier. In the same drawer under label

'*Cet. variegata*'

there is a specimen which is not *variegata*, but which corresponds with the descriptions of *Cetonia smaragdula* given by Fabricius and Olivier; and it closely resembles modern examples of the species in the British Museum. This specimen is one of the few beetles in Dr Hunter's Collection which have the elytra and the wings spread out, and that feature has led to its further recognition as a pre-Fabrician type of historical interest; it is, as Mr Arrow pointed out to me, the actual type specimen of Drury's *virens*. It was purchased by Drury at the sale of a collection belonging to a Mr Leman, and had evidently been acquired by Hunter, probably along with the Goliath beetle (see p. 87) and certain other specimens from Drury's Collection. Its identity is at once apparent when the excellent figure in Drury's work is compared with the type specimen.

In his earlier descriptions of this species, Fabricius makes no reference to Drury, and he states that his *smaragdula* was founded on a Hunter specimen; later, however, in *Ent. Syst.* I, 2, p. 134, No. 34, he refers to Drury. Olivier follows Fabricius and amplifies his description and he also refers to Hunter's Collection. Olivier's figure represents the insect with closed elytra and may have been drawn from another specimen in Hunter's Collection, before the acquisition of Drury's type; but I have not found another specimen (presumably the type of *smaragdula* Fab.).

Description of Type, *Scarabaeus virens* Drury. Form obovate; plano-convex above. Glossy green suffused with yellow; the antennae, the gula, and the abdomen dark reddish brown, the abdominal sterna each with a blood-red median spot, and a yellowish patch near the end of the pygidium. The body is very smooth, faintly punctured, and the underparts are very thinly fringed with light-coloured hairs.

The *head* is moderately large, oblong, and punctulate. The clypeus is transverse, roughly semicircular in shape, and the front edge is upturned; beside each pointed eye-ridge there is a small angular impressed line, otherwise the clypeus is continuous with the frons. The maxillae have each two teeth, one of which is thick and blunt.

The *pronotum* is triangular and convex and faintly punctulate; it is marginate only in part, an impressed border line in front arises near the angles and is continued about half-way along the sides. The front of the pronotum is narrowed and excavate, with the front angles prominent and sharp; the sides, which are very sloping, curve gently inwards to the rounded hind angles; the base is sinuate, with a wide arch in the middle between two small lobes.

The *scutellum* is very large and is triangular; but the sides are squared off near the convex base where it is sharply right-angled.

The *elytra* do not cover the abdomen entirely and are not marginate; the bases are slightly excavate and inturned, the outer margins are nearly parallel with the suture and are abruptly rounded in at the apex. The surface of the elytra is punctulate-striate; the first three primary costae are clearly defined by the deeply impressed bordering lines; the interval between the sutural and second costae is greater than that between the second and third or intra-humeral; the intervals are irregularly punctulate. The line of the sutural costa almost reaches the apical margin; the inner bordering line of the second costa is continued round the apex, forming a loop with the second-last lateral line. The lateral lines are irregular and partly obsolete. The functional wings are well developed.

The *pygidium* is strigose-reticulate. The front *coxae* are large and contiguous. The *prosternum* is not prominently produced. The middle coxae are separated by the *mesosternum*, which is produced as a long, curved and blunt-pointed horn-like process; this process extends in front of the junction of the front coxae, and it is marked by a median longitudinal impressed line. The hindmost *femora* are conspicuously broad and flattened, and have a row of five or six large punctures bearing hairs on the under surface, which is slightly punctulate. The front *tibiae* have each three toothlike projections (one very small) and a distal articulated spine. The middle and hind tibiae are semi-cylindrical; on the outer surfaces there are rows of large punctures bearing short bristles, and the distal ends (imperfect in this type specimen) show traces of a fringe of short spines and (at the inner edge) two larger spines.

Length 30 mm.; breadth (across the elytra) 16 mm.
Hab. America (Fab.), South America (Oliv.).

32. *Anomala lucicola* (Fab.)

Coleopterorum Catalogus, pars 66 (F. Ohaus, 1918), Scarabaeidae, Rutelinae, etc., p. 111.

North America.

Syn. *Melolontha lucicola* Fab., *Ent. Syst.* Supplement, p. 132, Nos. 66–67 (1798); *Syst. Eleuth.* II, p. 174, No. 85 (1801).
M. errans Oliv., *Ent.* I, 5, p. 45, pl. 8, fig. 92 (1789).

The type (*Melolontha errans* Oliv.) of this species is noted in the card-index of Dr Hunter's Collection as missing; but in Cabinet A, drawer 2, there are three specimens misplaced under label

'*Mel. arthritica*'

which are evidently co-types of *errans* Oliv. All three are males. Mr Arrow has examined these and he is satisfied as to their identity. He considers that Ohaus is wrong in regarding Olivier's *errans* as synonymous with *errans* Fab. (see Ohaus, *Col. Cat.* pars 66, p. 72), because Olivier's *errans* is clearly distinct from his (Olivier's) *praticola*, as is evident when his figures of the two species are compared. Mr Arrow regards *lucicola* Fab. as identical with the *errans* of Olivier.

Description of Co-type, *Melolontha errans* Oliv.
Form elongate-oval, convex, smooth and glossy, the underparts more or less thinly clothed with hair. Testaceous, with

the head (except the clypeus), the central area of the
pronotum, the scutellum, the margins of the elytra, the meso-
and meta-thoracic sterna black; the legs and the abdominal
sterna reddish brown.

The *head* is closely punctured; the clypeus is transverse,
straight in front, with the angles rounded and the edge upturned, and closely punctured; there is a distinct transverse
carina between the clypeus and the frons.

The *pronotum* is transverse, narrowed and excavate in
front, the sides slightly angular, the base sinuate with a slight
median lobe; the front angles are rounded, the hind angles
are sharp and are almost right angles; the surface of the pronotum is irregularly punctured, there is a small mid-lateral
prominence at each side and a large shallow depression at
each front angle. The pronotum is marginate; but the
marginal stria of the base is very fine and is incomplete. The
scutellum is broad, bluntly triangular and punctured.

The *elytra* are moderately short, marginate except at the
bases, the outer sides are nearly parallel with the suture, and
the apices are doubly rounded; the shoulder callus is very
prominent, and there are five deeply impressed and punctured dorsal striae; the interval between the sutural and the
second costa is the widest of the three and is irregularly punctate, the interval between the second and the intra-humeral
costa (which becomes obsolete towards the apex) is narrow
and is punctate near the base.

The *propygidium*, which is exposed, and the prominent *pygidium* are closely punctured. The front *coxae* are contiguous.
The *prosternum* is not produced behind the front coxae. The
mesosternum is produced as a short blunt process between the
middle coxae, and it is marked by a shallow oval furrow. The
legs are moderately long; the hindmost femora are broader
than those of the middle legs and are flattened; the front
tibiae have each two tooth-like projections; the middle and
hind tibiae are spinose, with the distal ends fringed with
short spines, and each tibia bears terminal spurs; the tarsi of

PLATE 20

Melolontha errans Oliv. × 6½

the middle legs are long and each has two simple claws, the longer one cleft; the other tarsi are imperfect in this type specimen.

Length 9 mm.; breadth (across the elytra) 5 mm.
Hab. not stated.
See Plate 20.

33. *Anomala vitis* (Fab.)

Coleopterorum Catalogus, pars 66 (F. Ohaus, 1918), Scarabaeidae, Rutelinae, etc., p. 78.

Middle and South Europe, England.

SYN. *Melolontha vitis* Fab., *Syst. Ent.* p. 37, No. 26 (1775); *Sp. Ins.* 1, p. 41, No. 34 (1781); *Mant. Ins.* 1, p. 21, No. 41 (1787); *Ent. Syst.* 1, 2, p. 167, No. 54 (1792); *Syst. Eleuth.* 11, p. 172, No. 69 (1801); Oliv. *Ent.* 1, 5, p. 34, pl. 2, figs. 12 *a* and *c* (1789).
M. holosericea Ill.

Two specimens under label

'*Mel. vitis*
Fabr. pag. 41, No. 34'

in Cabinet A, drawer 2, are evidently the examples on which Fabricius founded this species; they are exactly according to his description, and they answer closely to the description given by Olivier. There is also a close resemblance between these co-types and Olivier's figures of the species. I have compared them with a modern example of the species in the 'Bishop' Collection, and they match it exactly.

Description of Co-type, *Melolontha vitis* **Fab.** *Male.* Form oval and sub-cylindrical, broadening out a little beyond the shoulders, convex, very compact. Bright bronze-green, with the prothoracic episternum and the metasternum brilliant brassy coloured; the underside of the body thinly clothed with light hair.

The *head* very convex, closely punctured; the clypeus transverse, almost straight in front, with the angles a little rounded and the edge upturned; a slight transverse carina between the clypeus and the frons.

The *pronotum* is bluntly triangular and transverse, narrowed and excavate in front, the sides are rounded, the base is sinuate and slightly lobed in front of the scutellum, the front angles are sharp and the hind angles are rounded; the surface is closely punctured and is marked with irregular shallow depressions at the sides, a small median double depression at the front and a small roughly triangular median one at the base; the pronotum is marginate, but the marginal stria of the base is incomplete. The *scutellum* is broad, heart-shaped, and punctured.

The *elytra* are moderately short, broadened out behind, marginate on the outer sides which are nearly parallel with the suture, and have well-developed membranes; the apices are rounded, and the shoulder and apical callus are both prominent; there are five clearly impressed and punctured dorsal striae, the sutural stria extends to the apical margin, the interval between the sutural and the second costa is the broadest of the three and shows traces of interrupted lines, and all three intervals are irregularly punctate.

The *propygidium* is exposed and is closely punctured; the prominent *pygidium* is rugulose, and it is visible from beneath, the junction with the last ventral segment not being at the end of the body but a little beyond and ventral, as the specimen is a male. The front *coxae* are contiguous. The *prosternum* is not produced behind the front coxae. The *mesosternum* is produced as a small knob-like process immediately behind the middle coxae, and it is marked by a median longitudinal deeply impressed line. The *legs* of this type specimen are imperfect; the hindmost femora are broad and flattened; the front tibiae have each two tooth-like projections, the distal one long.

Length 16 mm.; breadth (across the elytra) 9 mm.
Hab. Europe and America (Fab. and Oliv.).

The *female* is similar to the male in size and in other respects, except that the pygidium is not visible from beneath, and the median longitudinal deeply impressed line on the mesosternum is not simple but is shaped like the head of a two-pronged arrow with the prongs outcurved.

34. *Anomala innuba* (Fab.)

Coleopterorum Catalogus, pars 66 (F. Ohaus, 1918), Scarabaeidae, Rutelinae, etc., p. 100.

United States of America (Middle and South).

SYN. *Melolontha innuba* Fab., *Mant. Ins.* 1, p. 22, No. 45 (1787); *Ent. Syst.* 1, 2, p. 169, No. 61 (1792); *Syst. Eleuth.* 11, p. 173, No. 78 (1801); Oliv. *Ent.* 1, 5, p. 46, pl. 8, fig. 93 (1789). *Anomala minuta* G. Horn (nec Burm.). *A. rufiventris* Dej.

Two specimens in Cabinet A, drawer 2, under label

'*Mel. innuba*
Fabr. MSS'

are not that species; they are different insects, namely, *Valgus hemipterus* and *Hoplia philanthus*. In the same drawer there are two specimens under label

'*Mel. sylvicola*'

and one of these (the left-hand one) answers the descriptions of *innuba* given by Fabricius and Olivier; it also corresponds with Olivier's figure and it matches exactly a British Museum example with which it has been compared; it is evidently the misplaced type.

Description of Type, *Melolontha innuba* Fab. Form oval, broadening out slightly beyond the shoulders, plano-convex, very compact. Glossy reddish black above, with the sides of the pronotum testaceous; the meso- and meta-sterna black, the abdominal sterna reddish brown; the antennae, the underparts of the head, the prothoracic episterna and sternum, the pygidium, and the coxae and femora of the legs testaceous. The underparts very slightly hairy.

The *head* closely punctured, flattened, with three shallow impressions, a median one on the frons and one on each side of the clypeus; a distinct transverse carina (interrupted in the middle) between frons and clypeus; the clypeus transverse and semicircular, with its front edge upturned; the clubs of the antennae long, the eyes large, the eye-ridges curved and pointed.

The *pronotum* is bluntly triangular and transverse, narrowed and excavate in front, the sides are a little angular, and the base is sinuate, slightly lobed in front of the scutellum; the front angles are sharp, the hind angles are a little rounded, and the surface is closely punctured; the pronotum is marginate, the front margin is interrupted in the middle, but the marginal stria of the base is complete. The *scutellum* is broad, rounded, and punctured.

The *elytra*, which do not cover the abdomen entirely, are a little dilated about the middle, are marginate on the inner and outer sides and have membranes; the apices are rounded in at the suture and the shoulder callus is prominent; the costae are slightly elevated, there are five clearly impressed and punctured dorsal striae, the sutural stria extends to the apical margin; the interval between the sutural and the second costa is the broadest one, and it is marked by two lines of punctures, which coalesce towards the apex, and between these there is a half-way line of punctures; the striae of the second costa are continuous with two of the lateral striae at the apex, forming a sharp loop; the striae of the intra-humeral costa become obsolete towards the apex; there is a line of punctures in the second interval.

The prominent *pygidium* is rugulose. The front *coxae* are contiguous. The *mesosternum* is produced as a very short process behind the middle coxae, and it is marked by a median longitudinal impressed line. The hindmost *femora* are moderately broad and flattened. The front *tibiae* have each two tooth-like projections, the distal one long; the middle tibiae are spinose and have two long distal spines; the middle *tarsi* have each two long and curved simple claws, the outer one is divided into two; the hind tibiae have two oblique rows

PLATE 21

Melolontha innuba Fab. × 8

of short spines on the outer side, the distal ends are fringed with short spines and bear (at the inner edge) two long spines.

Length 7 mm.; breadth (across the elytra) 4 mm.
Hab. not stated.
See Plate 21.

35. *Popillia rufipes* (Fab.)

Coleopterorum Catalogus, pars 66 (F. Ohaus, 1918), Scarabaeidae, Rutelinae, etc., p. 147.
Sierra Leone.

SYN. *Cetonia 4-punctata* Fab., *Mant. Ins.* I, p. 27, No. 12 (1787).
   C. rufipes Fab., *Ent. Syst.* I, 2, p. 129, No. 19 (1792); *Syst. Eleuth.* II, p. 139, No. 21 (1801).
   C. quadripunctata Oliv., *Ent.* I, 6, p. 80, pl. 10, fig. 93 (1789).

The original *rufipes* Fab. (*Mant. Ins.* I, 1787, p. 27, No. 7) is apparently not this West African Ruteline, but is the South African Cetoniine now known as *Ischnostoma cuspidata* Fab.

Fabricius also, unfortunately, misnamed this insect 4-*punctata* (*Mant. Ins.* I, 1787, p. 27, No. 12), he having already (*Sp. Ins.* I, 1781, p. 52, No. 8) applied that name to a different species, namely, the South European *Potosia morio* Fab. ab. *quadripunctata* Fab.

The specimen in Cabinet A, drawer 2, under label

'*Cetonia 4-punctata*'

is not that species; but in the same drawer there is an unlabelled specimen which answers the descriptions given by Fabricius and Olivier, and Olivier's figure closely resembles it. It has been examined by Mr Arrow, and it is evidently the type of the misnamed species 4-*punctata*.

Description of Type, *Cetonia 4-punctata* Fab. The specimen is a *male*. Form oval, broadest across the shoulders, plano-convex, the elytra short and truncate, the propygidium and pygidium visible from above. Dull glossy black above, with two whitish grey spots on the pygidium; shiny greenish black beneath and clothed with fringes of light grey hair; the antennae and the legs reddish testaceous.

The *head* flattened, punctulate and rugulose; a slightly impressed line between the frons and the clypeus; the clypeus transverse and with the front edge upturned; the clubs of the antennae moderately long; the eyes large, the eye-ridges short, curved and pointed.

The *pronotum* is convex and roughly octagonal in shape, a little narrowed and excavate in front; the sides are angular and marginate, the base is angular and is arcuate in front of the scutellum; the front and hind angles are sharp; the surface is closely punctulate, with small oval punctures and with rugulose effect, except about the base which is smooth. The *scutellum* is triangular and is marked anteriorly with linear punctures.

The *elytra*, which are short and truncate and do not cover the abdomen entirely, are narrowed a little towards the apices, there abruptly rounded in, and are marginate; the outer margin is deflexed at the shoulder, the base is sinuate, the shoulder callus is prominent, and at each sutural angle there is a small apical spine; the surface is irregular and is marked by slight and irregular costae and irregular rows of small punctures (punctate-striate); in the wide interval between the sutural and the second costa there is an irregular row of larger punctures which are faintly impressed and not close together; about the shoulders, where the costae and striae are incomplete, the surface is punctate.

The *propygidium* is punctulate. The prominent *pygidium* is punctulate and is visible from beneath, the junction with the last ventral segment not being at the end of the body but beyond and ventral, as the specimen is a male. On the pygidium there are two wide apart and conspicuous oval spots or patches of whitish grey hairs. The front *coxae* are contiguous. The *metasternum* is produced as a prominent blunt-pointed process reaching the front coxae. The *legs* are stout, with stout tibiae and tarsi; and each tarsus has two

PLATE 22

Cetonia 4-punctata Fab. ♂ × 6

long, hook-like, and unequal claws. The hindmost femora are broad and flattened. The front tibiae have each two distal tooth-like projections, which are short, sharp and conical; the outer claw of each front tarsus has a small, thin and sharp tooth near the tip. The middle tibiae have one oblique row of short spines on the outer side and the hind tibiae have two rows.

Length 15 mm.; breadth (across the elytra) 10 mm.
Hab. India (Fab.), East Indies (Oliv.).
See Plate 22.

Sub-family DYNASTINAE

36. *Enema infundibulum* Burm.
(= *Enema pan* Fab. ♂ var.)

Burm. *Handb.* v, p. 234. *Catalogus Coleopterorum* (Gemminger and Harold, 1869), IV, Scarabaeidae, p. 1263.

Central and South America.

SYN. *Scarabaeus enema* Fab., *Mant. Ins.* I, p. 4, No. 12 (1787); *Ent. Syst.* I, 1, p. 6, No. 12 (1792); Oliv. *Ent.* I, 3, p. 22, pl. 12, fig. 114 and pl. 17, fig. 157 (1789).
Geotrupes enema Fab., *Syst. Eleuth.* I, p. 6, No. 13 (1801).

There is no example in Dr Hunter's Collection labelled *Sc. enema*; probably the original label has been lost or destroyed. There is, however, in Cabinet C, drawer 32, a specimen (not labelled) which agrees perfectly with the descriptions of *enema* given by Fabricius and Olivier, and Olivier's figure closely resembles it. I have also carefully compared this specimen with a modern example of the species in the British Museum, and I think it is undoubtedly the type. According to Fabricius it came from India, but that is a mistake; Olivier says it is found in Brazil.

Description of Type, *Scarabaeus enema* **Fab.** (= isolated *male* form of *Enema pan* Fab.). Form robust, ovate and moderately convex; dull glossy black above, bright glossy reddish brown beneath and with thin fringes of short tawny hair. The *clypeus* narrow and tapered, slightly emarginate, its surface punctate. The canthus or brow-ridge produced as a short process in front of the eye. The *mandibles* bilobed, strongly bifid; the *maxillary palps* four-jointed (the first joint triangular, the second about three times the length of the first, the third cylindrical, the fourth elongate-ovate, truncate at the tip, and almost equal in length to the second plus the third); the *labium* roughly triangular, with the ligula bluntly conical and the surface deeply pitted and clothed with reddish hairs; the *labial palps* short and three-jointed (the first two joints obconical, the third longer and ovate). On the *head* a long and stout horn, recurved towards the end and there flattened out and deeply notched, a bifid termination with blunt tips; this horn directed upwards and forwards, convex and punctate in front, convex and rugose behind, each side flattened and with a shallow excavation extending about half an inch from the base, behind which there is a tuft of short hairs.

The *pronotum* strongly marginate all round, except towards the front angles; the front narrow, the base trisinuate, the hind angles sharp, the sides rounded and very rugose; on each side margin, and a little in front of the hind angles, a slight notch where the sides widen out to about as far as the middle and from there converge towards the head. The posterior portion of the pronotum raised in the middle into a prominent horseshoe-shaped mass or hump, which is mainly smooth and glossy and which slopes gradually down to the narrow front; a further elevation of the hump posteriorly forms a long, stout and simple thoracic horn which is punctate and angular (flattened on its four surfaces), except towards the pointed tip, and which curves outwards. In side view the outward and arching bend of this horn, along with the slope from its base to the head, represents a shapely half-circle. The *scutellum* short and bluntly triangular (heart-shaped); its surface rugose, except about the margin which is smooth. The front portion of the *metasternum* punctate, more

PLATE 23

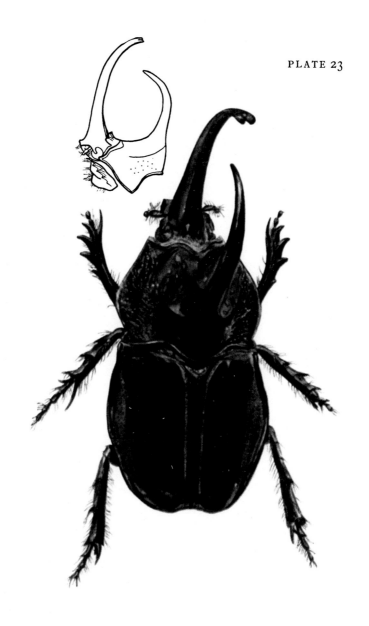

Scarabaeus enema Fab. × 1½ (**Enema pan** (Fab.) ♂ var.)
and side view of the horns

or less closely set with irregular oval punctures bearing short and fine hairs; elsewhere the surface of the metasternum is smooth and shining (glabrous).

The *elytra*, which are broadened out a little about the middle, are in great part faintly striate and irregularly punctulate; the outer sides strongly marginate and conspicuously pitted with large, round and deep punctures; along each side of the suture a deeply impressed line of confluent punctures, the punctures becoming distinctly separate towards the apices of the elytra.

The *abdominal sterna* glossy and more or less punctulate; three of the segments have each a row of large and irregular punctures bearing short tawny hairs. The *pygidium* closely punctate anteriorly, the punctures bearing hairs which form a thick tawny fringe.

The front *tibiae* quadridentate, having each four tooth-like projections, the first or proximal one very small and the fourth smaller than the second and third; each front tibia has a sharp terminal articulated spine, and the middle and hind tibiae have each two terminal articulated spines.

Length 55 mm.; breadth (across the elytra) 30 mm.
Hab. India (Fab.), Brazil (Oliv.).
See Plate 23.

37. *Enema pan* (Fab.)

Catalogus Coleopterorum (Gemminger and Harold, 1869), IV, Scarabaeidae, p. 1263.

Brazil.

SYN. *Scarabaeus pan* Fab., *Syst. Ent.* p. 5, No. 8 (1775).
♀ = *Scarabaeus quadrispinosus* Fab., *Sp. Ins.* I, p. 11, No. 36 (1781); *Mant. Ins.* I, p. 6, No. 38 (1787); *Ent. Syst.* I, 1, p. 15, No. 41 (1792); Oliv. *Ent.* I, 3, p. 33, pl. 19, fig. 179 (1789).
Geotrupes quadrispinosus Fab., *Syst. Eleuth.* I, p. 14, No. 43 (1801).

There is no example in Dr Hunter's Collection labelled *Sc. quadrispinosus*, so this is probably another instance of the original label having been lost or destroyed. There are, however, two specimens (not labelled) in Cabinet C, one of which (in drawer 32) is certainly the type; it closely agrees with the descriptions given by Fabricius and Olivier, and it exactly resembles Olivier's figure. The other specimen (in drawer 33) closely approaches the type.

Description of Type, *Scarabaeus quadrispinosus* Fab. (=*female* of *Enema pan* Fab.). It is, as Olivier says, about the size of *Oryctes rhinoceros* but a little broader, and very like that species in form and general appearance. Glossy dark chestnut-brown above, lighter beneath. The *head* has a long, stout and simple horn, recurved and slightly elbowed, with the base broad, the sides compressed and the tip blunt; the surface of this horn is rugose, there is a very slight indication of a tooth-like projection on the posterior edge and a small tuft of tawny hairs behind the base. The *clypeus* tapered, narrow, truncate, distinctly marginate and rugose; in front emarginate with a small tooth-like projection at each side (bidentate). The canthus or brow-ridge produced backwards as a horizontal process extending almost half-way across the middle of the eye, which is light reddish brown.

The *pronotum* transverse, strongly marginate all round, its surface very rugose, except about the middle of the sides where it is smooth; the base very trisinuate, the hind angles sharp, the sides rounded and converging towards the front which is excavated and trisinuate with the front angles produced as pointed processes. The anterior and greater part of the pronotum flattened about the middle and there insunk, forming a central hollow which is almost circular in outline; the posterior part strongly elevated into a prominent and arched hump, at the front of which (on the posterior part of the circular border of the anterior hollow) there are four small tubercle-like prominences, the inner or intermediate two larger and a little in advance of the outer two. The *scutellum* short and bluntly triangular; its surface rugose, except about the margin which is smooth. The *thoracic sterna* punctate and hairy.

The *elytra* punctate-striate at the sides anteriorly; about the middle a few faintly impressed striae, and along each side of the suture a deeply impressed and wavy stria; the outer sides strongly marginate, the margins upturned and sinuate and bent round at the shoulder angles towards the prominent callus.

The *abdominal sterna* feebly punctate, slightly rugose and fringed with short hairs. The *propygidium* not prominent; the *pygidium* protuberant and transversely sulcate, its surface punctulate and anteriorly a close fringe of long and tawny hairs.

The *femora* are rusty brown; the front *tibiae* are quadridentate, having each four tooth-like projections (the proximal one very small) and a terminal articulated spine; the middle tibiae have each one and the hind tibiae have each two terminal articulated spines.

Length 45 mm.; breadth (across the elytra) 26 mm.
Hab. India (Fab. in *Sp. Ins.*), East Indies (Oliv.), Cayenne (Fab. in *Syst. Eleuth.*).

The other specimen in Cabinet C, drawer 33, is elongate-cylindrical in form, dark chestnut-brown above, lighter beneath. The horn of the *head* is abruptly elbowed near the middle, its blunt tip is slightly enlarged and noticeably inturned, its sides are somewhat hollowed in an oblique direction behind so that it is biconcave latero-posteriorly.

The *pronotum* is strongly marginate all round; its surface rugose in front and smooth at the sides and behind, except about the lesser elevations where it is rugose; the base is trisinuate and the hind angles are rounded; the sides are slightly rounded and diverging a little from the base to beyond the middle, there becoming less prominent, concave and abruptly narrowed, tapering in towards the head; the front angles are sharp; the front edge of the high and arched posterior hump is rounded and has two small and sharp tubercle-like prominences, one on each side of the middle line and about 4 mm. apart; about 5 mm. beyond each of these intermediate prominences, and on the forward extension of the front edge, are indications of other two elevations, scarcely recognisable unless viewed from the side.

The *abdominal sterna* are punctate and shining. The *pro-pygidium* is prominent and rugulose; the *pygidium* is punctate, with a dense tuft of tawny hairs about the apex. Otherwise the description of this insect is the same as that of the type.

Length 44 mm.; breadth (across the elytra) 21 mm.

38. *Strategus titanus* (Fab.)

Catalogus Coleopterorum (Gemminger and Harold, 1869), IV, Scarabaeidae, p. 1264.

Jamaica.

SYN. *Scarabaeus aenobarbus* Fab., *Syst. Ent.* p. 10, No. 28 (1775); *Sp. Ins.* I, p. 10, No. 32 (1781); *Ent. Syst.* I, 1, p. 13, No. 37 (1792); Oliv. *Ent.* I, 3, p. 28, pl. 16, figs. 147 *a* and *b* (1789).

Geotrupes aenobarbus Fab., *Syst. Eleuth.* I, p. 13, No. 40 (1801).

♀ = *Scarabaeus eurytus* Fab., *Syst. Ent.* p. 7, No. 13 (1775); *Sp. Ins.* I, p. 7, No. 17 (1781); *Mant. Ins.* I, p. 5, No. 18 (1787).

S. titanus Fab., *Syst. Ent.* p. 10, No. 27 (1775).

S. simson Linn., Drury, *Illus. Exot. Ins.* I, p. 81, pl. XXXVI, figs. 3 and 4 (1770).

The specimen under label

'*Sc. aenobarbus*
Fabr. pag. 10, No. 32'

in Cabinet A, drawer 1, is not the one described as *aenobarbus* by Fabricius. In the same drawer there are two specimens under label

'*Sc. cupreus*
Fabr. pag. 32, No. 146'

which are not *cupreus*; but they closely correspond with the descriptions of *aenobarbus* and *eurytus* given by Fabricius and are evidently these misplaced types. Mr Arrow confirms this.

Fabricius described *aenobarbus* four times. His three later descriptions are repetitions of the first; and the only difference is that in the first one he describes the two small elevations of the clypeus as 'blunt', whereas in his later descriptions he refers to them as 'sharp'.

When Olivier examined Dr Hunter's Collection, he found that there were two examples of *aenobarbus* and that the smaller one had been described by Fabricius as a different species, namely, *eurytus*. Olivier considered *eurytus* to be the *female* form of *aenobarbus*, and he pointed out that it only differs from the male in having shorter thoracic horns. Olivier also mentioned that the thorax of the male varies. (He had seen another example, one given to him by M. Francillon.) Fabricius in his *Entomologia Systematica* abolishes his *eurytus* as a distinct species and, in agreement with Olivier, refers to it as the *female* form of *aenobarbus*.

The descriptions of *aenobarbus* by Fabricius and Olivier are very similar, except that Fabricius describes it as dull glossy black; Olivier says blackish brown above and brown beneath.

The two examples of *aenobarbus* agree well with Olivier's description and also resemble his figures, the only difference being that the larger one is (now) pitchy black.

Scarabaeus titanus Fab. is the same insect in its fully developed form; therefore the two Hunterian examples (the type of *aenobarbus* Fab. and the type of *eurytus* Fab.) are to be regarded as minor forms of *titanus* Fab. We further find that both are *small males*.

Description of Type, *Scarabaeus aenobarbus* Fab. = *Scarabaeus titanus* Fab. minor form of male. Form short, oval and convex; pitchy black, the underparts glossy and clothed with moderately long tawny hairs. The *clypeus* narrow and truncate, slightly reflexed, and rugose, except at the base; with a slight median transverse ridge upon which there are two very small tubercles. The canthus or brow-ridge produced backwards as a short process extending horizontally less than half-way across the middle of the eye, which is reddish brown.

The *pronotum* transverse, its breadth greater than its length, distinctly marginate all round, irregularly punctulate towards the sides but elsewhere smooth and shining; the base slightly trisinuate, the hind angles a little rounded or blunt, the sides diverging slightly from the base towards the middle and from there converging towards the narrow front, the front angles sharp. The anterior half of the pronotum depressed and hollowed, except at the narrowed front which is raised into a horn about as long as the head, recurved and with a recumbent incline, rectangular about the base, and flattened and notched (bifid) at the tip; the posterior and wider half elevated in front into two short tubercle-like prominences with a well-marked crescentic linear space between them. The *scutellum* small and triangular. The *thoracic sterna* closely punctulate and set with tawny hairs; the *metasternum* closely punctulate and set with tawny hairs anteriorly, smooth and shining posteriorly.

The *elytra* with a few feebly impressed striae, and with a distinct crenulate line (like a hem-stitch) along each side of the suture, and sparsely punctulate.

The *abdominal sterna* have each a row of punctures with projecting hairs; but the rows are interrupted and the punctures become obsolete on the median portions of the sterna where the surface is smooth and shining. The *pygidium* rounded, smooth and shining, slightly punctulate at the sides, and with a fringe of tawny hairs anteriorly.

The *legs* fringed with tawny hairs; the front *tibiae* quadridentate, having each four tooth-like projections (the proximal one very small), and with a conspicuous terminal articulated spine; the middle tibiae are shorter than the others, and each has a terminal articulated spine; the hind tibiae have each two terminal articulated spines.

Length 32 mm.; breadth (across the elytra) 17 mm.
Hab. America (Fab.), South America, in Jamaica (Oliv.).
See Plate 24.

Description of Type, *Scarabaeus eurytus* Fab.

PLATE 24

Scarabaeus aenobarbus Fab. × 3
(*Scarabaeus titanus* Fab. ♂ var.)
and side view of head and pronotum showing
configuration of the anterior horn and
one of the posterior prominences

(*Scarabaeus aenobarbus* Fab. ♀) = *Scarabaeus titanus* Fab. minor form of male. Brownish black. The tubercles of the *clypeus* very minute. The *pronotum* rugulose in front and closely punctate about the sides, the punctures oval and mostly contiguous. The anterior part of the pronotum not so markedly depressed and hollowed; the horn very short, less than half the length of the head, and distinctly notched (bifid) at the tip; the posterior part less elevated and the two tubercle-like prominences very slight, just recognisable. Otherwise the same as the preceding type, but smaller in size.

Length 30 mm.; breadth (across the elytra) 15 mm.

39. *Golofa hastata* (Fab.)

Catalogus Coleopterorum (Gemminger and Harold, 1869), IV, Scarabaeidae, p. 1265.

South America (Guiana).

SYN. *Scarabaeus hastatus* Fab., *Sp. Ins.* I, p. 6, No. 11 (1781); *Ent. Syst.* I, 1, p. 6, No. 11 (1792); Oliv. *Ent.* I, 3, p. 21, pl. 19, fig. 175 (1789).
    Geotrupes hastatus Fab., *Syst. Eleuth.* I, p. 6, No. 12 (1801).

The specimen under label

' *Sc. hastatus*
Fabr. pag. 6, No. 11 '

in Cabinet A, drawer 1, is not the one described as *hastatus* by Fabricius; it does not answer his description nor does it agree with the description given by Olivier, and it does not resemble Olivier's figure.

There is, however, in Cabinet C, drawer 33, a specimen (not labelled) which answers the description given by Fabricius, except that it is not black but of a dark brown colour. This specimen agrees also with Olivier's description, and his figure of *hastatus* closely resembles it. I have little doubt that it is the misplaced type. Mr Arrow has examined it and he confirms the identity.

Apparently Fabricius made two mistakes concerning this

type, he called it black and he said it was in the Banks Collection. It is known that Olivier had access to the Banks Collection and that he examined the types; but in his description of *hastatus* he says, "In specimine á me viso (du Cabinet de feu M. Hunter) color castaneus, magis aut minus obscurus". Fabricius, in his later work, the *Entomologia Systematica*, has corrected his error in location, for therein we read "Mus. Dr Hunter" instead of "Mus. Dom. Banks".

Olivier also says, "Cet insecte n'est peut-être qu'une variété du Scarabé porte-clef". But it is not *Golofa claviger* in Gemminger and Harold's *Catalogus Coleopterorum*. Among the numerous specimens of *claviger* in the British Museum no example approaching it can be found.

As it is clearly not the beetle commonly called *claviger*, it must meanwhile therefore be regarded as a distinct species of *Golofa*, namely, *Golofa hastata* Fab.

Description of Type, *Scarabaeus hastatus* **Fab.** Form elongate-cylindrical. Head and pronotum dark mahogany-brown, the elytra chestnut-brown; dark reddish brown beneath, except the femora which are light chestnut-brown; the under parts of the head and thorax clothed with yellowish hairs.

The *head* is armed with a long and simple horn, sharply pointed and strongly recurved. The *clypeus* is narrow and emarginate (bidentate, according to Olivier), with the angles sharp and the surface rugose. The canthus or brow-ridge is produced backwards as a short process extending horizontally half-way across the middle of the eye, which is yellowish.

The *pronotum* is almost as long as broad, distinctly marginate all round, its surface granulate and rugulose (finely punctate, according to Fabricius and Olivier) with punctures at the sides of the horn and with a distinct shallow impression; the base trisinuate and the hind angles are slightly blunted; the sides, gently rounded, diverge from the base towards the middle and from there narrow in towards the front; the front angles are sharp. The anterior half of the

PLATE 25

Scarabaeus hastatus Fab. × 2
and side view of the horns

pronotum is depressed; the posterior half is elevated in front into a short and thick median horn which is arched, convex above, concave and hairy beneath and trilobed at the apex (suggestive of a blunted spear-head). This horn has a recumbent incline, and its width about the base is 4 mm. The *scutellum* is small and triangular. The *metasternum* is slightly rugose in front and smooth behind.

The *elytra* are transversely rugose about the base, elsewhere finely granulate and rugulose and with some feeble irregular punctures; along each side of the suture there is a stria which is strongly impressed anteriorly. The colour of the elytra is a lighter shade of brown than that of the thorax.

The *abdominal sterna* are rugosely punctulate, except the median portions which are smooth and glossy. The *propygidium* is prominent; the *pygidium* is punctulate.

The *femora* are light chestnut-brown; the front *tibiae* are tridentate, having each three tooth-like projections and a small terminal articulated spine; the hind tibiae have each two terminal articulated spines. The middle legs have been broken off and lost.

Length 48 mm.; breadth (across the elytra) 24 mm.
Hab. South America.
See Plate 25.

40. *Xylotrupes gideon* var. *oromedon* Drury
= small *male* form of *X. gideon* (Linn.)

Catalogus Coleopterorum (Gemminger and Harold, 1869), IV, Scarabaeidae, p. 1267. *The Fauna of British India*, Coleoptera, Lamellicornia, Part I (Cetoniinae and Dynastinae), by G. J. Arrow, 1910, p. 262.

East Indies, India, China.

SYN. *Scarabaeus oromedon* Drury, *Illus. Nat. Hist.* I, p. 81, pl. XXXVI, fig. 5 (1770); Voet, *Cat. Syst. Coll.* La Haye, pl. 13, fig, 102 (1766); Fab. *Syst. Ent.* p. 4, No. 3 (1775); *Sp. Ins.* I, p. 5, No. 5 (1781); *Mant. Ins.* I, p. 4, No. 5 (1787); *Ent. Syst.* I, 1, p. 4, No. 4 (1792); Oliv. *Ent.* I, 3, p. 17, pl. 18, fig. 165 (1789).

Geotrupes oromedon Drury, Fab. *Syst. Eleuth.* I, p. 4, No. 4 (1801).

Olivier described and figured an example of this insect in Dr Hunter's Collection; it is noted in the card-index as missing, but I have been able to trace it. In Cabinet C, drawer 32, there are two specimens of *oromedon*; both are males of intermediate development. The smaller specimen, which has the wings spread out, is probably the example described and figured by Olivier. The larger one may be the Drury specimen, it corresponds exactly with Drury's figure, and it also corresponds with Olivier's figure of *phorbanta* Oliv.

The tooth on the upper edge of the cephalic horn is present in both; the thoracic horn of the smaller specimen does not extend as far as the cephalic horn and is bifurcate, but the points are short, not widely outspread. The colour of these examples is uniformly dark chestnut-brown; the head, the pronotum, and the legs are of a slightly darker shade. In other respects the description is exactly that of the typical species, *Xylotrupes gideon*, as given by Mr Gilbert J. Arrow in *The Fauna of British India*, Coleoptera, Lamellicornia, Part I (Cetoniinae and Dynastinae), 1910, pp. 262–4.

Length (of the smaller specimen) 41 mm.; breadth (across the elytra) 26 mm.

Length (of the larger specimen) 46 mm.; breadth (across the elytra) 28 mm.

Sub-family CETONIINAE

41. *Goliathus goliatus* (Drury)

Coleopterorum Catalogus, pars 72 (S. Schenkling, 1921), Scarabaeidae, Cetoniinae, p. 4.

West, Central and East Africa.

SYN. *Scarabaeus goliatus* Drury, *Illus. Exot. Ins.* I, p. 67, pl. XXXI (1770).

S. goliathus Linn., *Mant. Plant.* VI, p. 530 (1771).

S. goliatus Fab., *Syst. Ent.* p. 13, No. 41 (1775); *Sp. Ins.* I, p. 14, No. 51 (1781); *Mant. Ins.* I, p. 7, No. 54 (1787).

Cetonia goliathus Oliv., *Ent.* I, 6, p. 7, pl. 9, fig. 33 *c* (1789).

C. goliata Fab., *Ent. Syst.* I, 2, p. 124, No. 1 (1792); *Syst.*
Eleuth. II, p. 135, No. 1 (1801).
Goliathus giganteus Lam., *Hist. Nat. Anim. sans Vert.* IV,
p. 580 (1817).

As already stated (see *Deltochilum gibbosum*, p. 46) the
insects under label

'*Sc. goliatus*
Fabr. pag. 14, No. 51'

in Cabinet A, drawer 1, are neither of them that species. The
type specimen of *goliatus* is located in Cabinet C, drawer 33.
Considering the fact that it was obtained in or before the year
1767, its condition is remarkably perfect. It originally belonged to Drury and was first described by him; it is therefore
a Drury type and pre-Fabrician.

Concerning the history of this type, Drury gives the following note: "The specimen was brought from Africa by Mr
Ogilvie, Surgeon of His Majesty's ship the *Renown*, being
found floating, dead, in the river Gaboon, opposite Prince's
Island, near the equinoctial line". Westwood, in his revised
edition of Drury's work, 1837, p. 62, tells us more about it:

"Nearly seventy years", he says, "have elapsed since this insect
was first described, and yet the insect remains, as far as my knowledge extends, unique. It would appear that the specimen either
belonged to or passed into the hands of Dr Hunter after the death
of Mr Drury, for Fabricius describes the species with a citation of
the museum of Dr Hunter alone; and Olivier's figure was taken
from the specimen whilst it was in that gentleman's possession.
After his decease it passed, with his collection, by bequest, into the
possession of the University of Glasgow, where it now forms one
of the most interesting objects in the Hunterian Museum. Joseph
Hooker, Esqr., son of Sir William Hooker, the highly distinguished
botanist of Glasgow, tells me that the individual in question was
picked up by a sailor in the river above mentioned, and that it is
stated in the MSS of Dr Hunter that it cost Mr Drury £10".

Westwood also states that

In the Catalogue of the Insects of Mr Drury, which were sold
by auction at the Natural History Sale Rooms in King Street,
Covent Garden (now occupied by Mr J. C. Stevens), on the 23rd

of May, 1805, and two following days, the 95th lot is described as
'Scarabaeus goliathus, *var*.' Whence it would appear that the insect
here figured was not in the possession of Mr Drury at his decease
and that he only possessed the insect figured in the 3rd volume of
these Illustrations, pl. 40, which evidently on the authority of
Fabricius he had regarded as *a variety only* of the specimen here
figured.

As to this other specimen (concerning which Drury said,
"but I judge it to be a different species"), Fabricius does not
mention it in any of his works; but Olivier describes and
figures it (*Ent.* 1, 6, p. 7, pl. 5, fig. 33) along with the type, and
he states it to be "in the Cabinet of Mr Hunter". If it ever
was in Dr Hunter's possession, it has evidently been removed.

The description of the type given by Fabricius in his *Entomologia Systematica* differs from his earlier (and repeated)
descriptions; but the difference is merely that of a slightly
amplified description of the same insect. The type answers to
the descriptions by Fabricius; and it also closely agrees with
the more extended descriptions given by Drury and Olivier.
Drury's figure of it, drawn by Moses Harris in 1767, is
a strikingly accurate representation; and the resemblance
between Olivier's figure and the type is very close.

Description of Type, *Scarabaeus goliatus* Drury.
The specimen is a *male*. Form robust, elongate, broadest
across the base of the elytra; with a sub-globose prothorax
widest at the middle and contracted in front and behind, its
angles slightly rounded, its base strongly convex with a short
(truncated) basal lobe. The upper surface covered with a red
plum bloom and with markings of a dull white (flesh-
coloured) scaly substance.

The *head* is black, its upper surface is almost entirely dull
white (flesh-coloured), its under surface is glossy black and
punctate. Arising from each side of the fronto-clypeus, and
above the insertions of the antennae, there are two short and
thin ear-like horns: the clypeus is further produced as a large
and thick projection with curved and hollowed-out sides and
with two broad and divergent terminal horns, which curve

upwards and outwards and which have blunted triangular expanded tips; a slight longitudinal ridge extends along the middle of the hollowed head, from the base to the extremity between the bases of the horns. The head is about three-fourths of an inch in length; its breadth at the base is half an inch. The eyes are black.

The *prothorax* is very convex and arched; the front part slopes deeply down towards the head, which is directed downwards and forwards. The front part of the pronotum together with the head forms an obtuse angle in relation to the rest of the body. The *pronotum* is strongly narrowed in front; the middle of its posterior margin forms a short broad truncated lobe; its side margins are flattened out and shelf-like; its front angles are slightly rounded and the front is trilobed, the central lobe broader and slightly in advance of the small side lobes and forming a prominent arched elevation. The pronotum is of a dark red plum colour, with the lateral borders dull white and with five narrow and irregular longitudinal dull white stripes; the outer two lateral stripes join the white lateral border a little beyond the middle; the outer margins are black. The *scutellum* is long and triangular, the length of the base is about half the length of the convex sides, and the apex is blunted; it is of a dark red plum colour with a white triangular mark in the middle. The scutellum is an embossment on an insunk area which has the form of an equilateral triangle; and which is yellowish white at the sides and is situated between the basal portions of the elytra.

The *elytra* almost entirely cover the abdomen; the sides are a little sinuate (widening out about the middle) and converge towards the narrowed apices; the apical angles are blunt. The elytra have a deep red plum bloom, and on each elytron there is an irregular transverse dull white strip-mark along the margin of the base as far as the shoulder callus; the surface has a slightly rugose sculpturing; the sutural margins diverge widely at the base and become markedly costate towards the apices; the lateral and apical margins are costate and have a short thick fringe of golden yellow hairs.

The *mesosternal epimera* are visible from above, and at a lower level than the dorsal surface, as triangular pieces at the bases of the elytra and are dark green, punctate and hairy. The *prothoracic episterna* are prominent rounded bulgings, dull white and irregularly punctate (the punctures bearing short hairs), except on the margins and near the prosterna where each has an irregular black blotch. The *metasternum* is large and is produced in front into a broad mitre-shaped process, which extends forward between and beyond the coxae of the middle pair of legs, and which is marked by a median longitudinal line continuous with a conspicuous median furrow on the metasternum. The surface of the metasternum is more or less punctate, the punctures bearing golden yellow hairs. The sternum of the first abdominal segment is narrow and is very convex in the middle and there produced forward into a sharp-pointed process. All the *abdominal sterna* are rugulose-punctate and have, at the sides, golden yellow hairs which form a conspicuous fringe bordering the under surface of the abdomen. The *pygidium* is prominent; it is rugulose-punctate, the punctures bearing very short hairs. The underside of the body is glossy greenish black.

The *legs* are glossy greenish black. The tibiae are without tooth-like processes on the outer edge (the specimen being a *male*). The femora and tibiae of the middle and hind pairs of legs are densely fringed with golden yellow hair. The tarsi of the front legs are longer than the tibiae, and are longer than the tarsi of the middle and hind legs.

Length 4¼ inches; breadth (across the elytra) 2 inches.
Hab. Africa, Guinea (Fab.), Sierra Leone (Oliv.).

42. *Ischnostoma cuspidata* (Fab.)

Coleopterorum Catalogus, pars 72 (S. Schenkling, 1921), Scarabaeidae, Cetoniinae, p. 79.

Cape Colony, South Africa.

SYN. ♂ = *Cetonia rufipes* Fab., *Mant. Ins.* I, p. 27, No. 7 (1787).
C. *cuspidata* Fab., *Mant. Ins.* I, p. 27, No. 8 (1787); *Ent.
Syst.* I, 2, p. 129, No. 16 (1792); *Syst. Eleuth.* II, p. 138,
No. 18 (1801).
♀ = C. *nobilis* Fab., *Mant. Ins.* I, p. 27, No. 9 (1787).

In his preliminary notice of the Fabrician Types in Dr
Hunter's Collection, Professor Graham Kerr has included
Cetonia rufipes Fab.; but the type of this species is stated by
Fabricius to be in the Lund Collection.

The specimen under label

'*Cet. rufipes*
Fabr. MSS'

in Cabinet A, drawer 2, is not that species and has evidently
been misplaced.

43. *Gymnetis spence* Gory and Perch.

Mon. Cét. p. 70, 335, pl. 67, fig. 5 (1833). *Coleopterorum Catalogus*, pars 72 (S. Schenkling, 1921), Scarabaeidae, Cetoniinae,
p. 98.

Jamaica.

SYN. *Cetonia tristis* Fab., *Syst. Ent.* p. 45, No. 10 (1775); *Sp. Ins.*
I, p. 53, No. 13 (1781); *Mant. Ins.* I, p. 28, No. 21 (1787);
Ent. Syst. I, 2, p. 133, No. 33 (1792); *Syst. Eleuth.* II,
p. 143, No. 43 (1801).

There are two specimens under label

'*Cet. tristis*
Fabr. pag. 53, No. 13'

in Cabinet A, drawer 2, but neither of them is that species.
Apparently this type is not now in Dr Hunter's Collection.

44. *Pachnoda histrio* (Fab.)

Coleopterorum Catalogus, pars 72 (S. Schenkling, 1921), Scarabaeidae, Cetoniinae, p. 280.

Arabia, Egypt.

SYN. *Cetonia histrio* Fab., *Syst. Ent.* p. 51, No. 39(1775); *Sp. Ins.*
  I, p. 60, No. 54 (1781); *Mant. Ins.* I, p. 31, No. 66 (1787);
  Ent. Syst. I, 2, p. 152, No. 93 (1792); *Syst. Eleuth.* II,
  p. 158, No. 118 (1801); Oliv. *Ent.* I, 6, p. 45, pl. 10, fig. 94
  (1789).

The specimen in Cabinet A, drawer 2, under label

'*Cet. histrio*
Fabr. pag. 60, No. 54'

does not correspond with the descriptions of *histrio* given by
Fabricius and Olivier. There is, however, in the same drawer
a specimen under label *Mel. punctata* which is not that insect,
but which answers the description of *histrio* as given by
Olivier, and which also resembles his figure. This specimen
is evidently the Olivier type; Olivier states that the *Cetonia*
which he described and figured as *histrio* was in Dr Hunter's
Collection.

The Fabrician type is stated to be in Forskahl's Collection;
and, although the original description of it mainly agrees with
Olivier's description, Mr Arrow thinks it is not the same
insect. I have seen a modern example of *histrio* Fab. in the
British Museum, and it is apparently not the *histrio* of Olivier.

Description of Type, *Cetonia histrio* Oliv. Form
elongate-oval, narrowed behind the prominent shoulders,
flattened above. Dull glossy testaceous and greenish black
above, with five whitish marginal markings on each elytron
and with five whitish spots on the pygidium; greenish black
beneath, with whitish spots on the abdominal sterna, and
with fringes of light-coloured hair.

The *head* is black, elongate and closely punctured, and
between the prominent eyes there are two long depressions
separated by a blunt longitudinal carina; the clypeus is a

PLATE 26

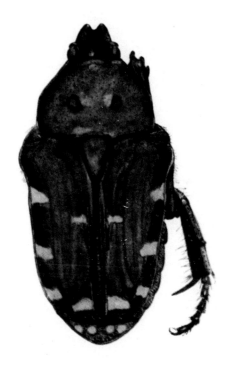

Cetonia histrio Oliv. $\times 5\frac{1}{4}$

little narrowed in front and is deeply notched, its front edge having the appearance of two slightly upturned teeth.

The *pronotum* is narrow in front, with a slight frontal lobe and sharp front angles; the base is broad, widely excavated in front of the scutellum, and the hind angles are very round; the sides are angular and slightly marginate; the surface is plano-convex, thinly punctate and testaceous with two large dark green semilunar spots in the middle of the disc. The *scutellum* is long, triangular and testaceous.

The *elytra* are flattened, shortened, and narrow (the dorso-lateral parts of the body are exposed); the bases are sinuate, the inner and outer sides are marginate, but the margins of the outer sides become obsolete towards the apices; the outer sides are strongly sinuate beyond the prominent shoulder callus, and are distinctly rugose; the apical angles are sharp and the apical callus is prominent; besides the sutural costa there are two dorsal costae which are continuous at the prominent apical callus and which together form a prominent raised and curved loop with an abrupt bend; these costae are bordered by rows of punctures, and the intervals are more or less closely punctate. The elytra are testaceous, broadly bordered with dark green, the border being interrupted by five whitish bar-markings (three on the outer side, one at the apical angle; and the smallest one, which is not mentioned by Olivier, near the middle of the inner or sutural side).

The angles between the prothorax and the shoulders of the elytra are occupied by the mesothoracic epimera; the meta-thoracic epimera, the hind coxae, and the side portions of the upper surface of the abdomen are also visible from above. The *propygidium* is partly exposed and is rugulose; the prominent *pygidium* is rugulose, covered with very short and fine hairs, and is marked with four round whitish spots. The *abdominal sterna* have each a transverse line of punctures and two large oval whitish spots situated one midway between each side and the middle line. The *mesosternum* is produced as a thick blunt process projecting forwards between and in front of the middle coxae; it is rugulose, except about the middle which is smooth and marked only by a median longitudinal impressed line and a few punctures. The front *coxae* are contiguous, ovate-conic and projecting; the front *tibiae* are ridged on the upper surface and have each three tooth-like projections and a terminal articulated spine; the hindmost *femora* are arched; the middle and the hind tibiae have each a small blunt spur on the outer side, are digitate distally, and have two terminal articulated spines; the tarsal claws are equal and simple.

Length 14 mm.; breadth (across the shoulders) 8 mm.
Hab. Egypt (Fab.).
See Plate 26.

45. *Oxythyrea haemorrhoidalis* (Fab.)

Coleopterorum Catalogus, pars 72 (S. Schenkling, 1921), Scarabaeidae, Cetoniinae, p. 330.

South Africa, Mozambique.

SYN. *Cetonia haemorrhoidalis* Fab., *Syst. Ent.* Appendix, p. 819, Nos. 34–35 (1775); *Sp. Ins.* I, p. 59, No. 48 (1781); *Mant. Ins.* I, p. 31, No. 58 (1787); *Ent. Syst.* I, 2, p. 148, No. 79 (1792); *Syst. Eleuth.* II, p. 154, No. 97 (1801); Oliv. *Ent.* I, 6, p. 55, pl. 4, fig. 24 and pl. II, fig. 24 *b* (1789).
Leucocelis haemorrhoidalis Fab.
Scarabaeus ruficollis De Geer.
Cetonia cruenta Gmelin.

This species is represented by three specimens under label

'*Cet. haemorrhoidalis*
Fabr. pag. 58, No. 49'

in Cabinet A, drawer 2, which closely correspond with the descriptions given by Fabricius and Olivier, and which resemble Olivier's figures drawn from specimens in the Gigot d'Orcy Collection. The example chosen as the type agrees perfectly with modern examples in the British Museum Collection.

Péringuey, in his description of "this most variable species" (*Trans. S. Afr. Phil. Soc.* XIII, 1907, pp. 477, 479, pl. 47, fig.

19), says that the base of the pronotum is "quite straight" in front of the scutellum; but in the Fabrician type specimen it is perceptibly emarginate. He describes the elytra as being "much attenuated behind"; the elytra of the type are only a little attenuated posteriorly. He further states that there is no distinct row of punctures along the outer margin of each elytron; but the type has a very distinct row of punctures along the outer margin. According to Péringuey the Fabrician type specimen would appear to belong to var. *viticollis*.

I have examined three modern examples of *haemorrhoidalis* Fab. in the 'Bishop' Collection, and have compared these with the type. One of them has the longitudinal double rows of irregular punctures on the elytra very clearly marked. With reference to Péringuey's description, in none of these 'Bishop' examples are the elytra "*much* attenuated behind", and all three have a distinct *duplicate* row of punctures along the outer margin of each elytron.

Description of Co-type, *Cetonia haemorrhoidalis* Fab. Form elongate and plano-convex, smooth and shining; crimson and black above with bright green elytra; glossy black beneath, with the prosternum and the last two abdominal sterna crimson; and with six small white spots along each side of the body (one on each mesothoracic epimeron, one on the sharp posterior angle of each hind coxa, and one on the posterior edge at each side of the first four abdominal sterna) and one small white spot at the apical margin of each elytron; the legs fringed and the underparts of the body thinly clothed with light reddish hair.

The *head* is black, elongate, closely punctured, and hollowed out along each side of the blunt longitudinal carina between the prominent eyes, which are divided in front by thin eye-ridges; the clypeus is a little narrowed in front and is deeply notched, its front edge is bidentate in appearance and upturned, and it is more finely and closely punctured than the frons.

The *pronotum*, dark crimson with the middle of the disc

black, is convex and punctured (not closely); in front it is narrow, excavate and sinuate, with a slight but distinct median lobe overarching the vertex of the head; its rounded sides are a little angular and are marginate, the front angles are sharp and the hind angles are rounded; its broad base is strongly rounded and is perceptibly emarginate in front of the scutellum. The *scutellum* is triangular, very sharply pointed, with concave sides, and upon its surface there are two or three punctures.

The *elytra*, bright green and with a squarish white spot at the middle of each apical edge, are shortened, leaving the pygidium and a small part of the propygidium exposed; the bases are angular; the outer sides are in greater part marginate, strongly sinuate behind the prominent shoulder callus and abruptly rounded in at the apices, which are almost straight with the apical angles sharp. The flattened dorsal surface of each elytron shows uneven furrows and elevations, and is marked by three double but irregular and imperfect rows of black punctures; the punctures of the two rows next the suture become confluent near the middle, consequently these rows are in greater part deeply impressed black lines; the same is true of the next two rows, but about the middle the inner one is broken aside and the outer row ends abruptly; the two intra-humeral rows become obsolete towards the prominent apical callus and end confluently as short broken lines; outside the intra-humeral double row there is a line of faint punctures, and along the outer margin a very distinct row of punctures extends from the shoulder to the apex, where the dorsal double rows appear to be continuous (forming loops) with this outer row and other faintly indicated lateral rows; examination with a lens of twenty diameters magnification reveals fine punctures thinly scattered throughout the spaces between the rows.

The *meso-* and *meta-thoracic epimera* and the hind *coxae* are visible from above. The prominent *pygidium* is light crimson colour, and is a little rugulose, with incomplete ring

PLATE 27

Cetonia haemorrhoidalis Fab. × 6½

and half-ring punctures. The *mesosternum* is partly smooth and partly rugulose and punctate, most of the punctures bearing short hairs; it has a median impressed line, and it is produced as a short and thick knob-like process projecting between and slightly in front of the middle coxae. The front *tibiae* are finely ridged, and are sharply bidentate, with a small distal articulated spine; the middle tibiae have a slight spur at the middle, and are distally digitate, with two articulated spines; the hindmost *legs* are long, the femora are broad and arched, the tibiae have a slight spur near the middle and are distally digitate and have two sharp articulated spines, one longer than the other; the tarsal claws are equal, curved, simple and wide apart.

Length 11 mm.; breadth (across the shoulders) 6 mm.
Hab. Cape of Good Hope (Fab.).
See Plate 27.

46. *Genuchus hottentottus* (Fab.)

Coleopterorum Catalogus, pars 72 (S. Schenkling, 1921), Scarabaeidae, Cetoniinae, p. 373.

Cape Colony, Transvaal.

Syn. *Cetonia cruenta* Fab., *Mant. Ins.* I, p. 32, No. 69 (1787); *Ent. Syst.* I, 2, p. 153, No. 97 (1792); *Syst. Eleuth.* II, p. 159, No. 124 (1801); Oliv. *Ent.* I, 6, p. 57, pl. 6, fig. 37 and pl. 7, fig. 58 (1789).

In his preliminary notice of the Fabrician Types in Dr Hunter's Collection, Professor Graham Kerr has included *Cetonia cruenta* Fab.; but the type of this species is in the Banks Collection (British Museum, London), as stated by Fabricius and Olivier.

The specimen in Cabinet A, drawer 2, under label

'*Cet. cruenta*
Fabr. MSS'

is not the insect described as *cruenta* by Fabricius and Olivier.

47. *Cetonia 'erdenta'*

In his preliminary notice of the Fabrician Types in Dr
Hunter's Collection, Professor Graham Kerr mentions the
following: "*Cetonia erdenta* Fabr., MSS. One specimen".
There is no specimen under that name in Hunter's Collection,
and '*erdenta*' does not occur in the works of Fabricius and
Olivier; I have also failed to trace it in the Catalogues.
Probably the label *Cet. cruenta* was misread '*erdenta*' when
the card-index of the Hunterian Collection was prepared.

Sub-family TRICHIINAE

48. *Trichiotinus bibens* (Fab.)

Coleopterorum Catalogus, pars 75 (S. Schenkling, 1922), Scarabaeidae, Trichiinae, etc., p. 24.

North America.

SYN. *Trichius bibens* Fab., *Syst. Ent.* p. 40, No. 3 (1775); *Sp. Ins.*
I, p. 48, No. 3 (1781); *Mant. Ins.* I, p. 26, No. 5 (1787);
Ent. Syst. I, 2, p. 121, No. 8 (1792); *Syst. Eleuth.* II,
p. 132, No. 8 (1801).
Cetonia bidens Oliv., *Ent.* I, 6, p. 62, pl. 10, fig. 87 (1789).

Two specimens under label

'*Tr. bibens*
Fabr. pag. 48, No. 3'

in Cabinet A, drawer 2, are evidently this species, they
answer the descriptions given by Fabricius and Olivier. One
of them is apparently the Hunterian insect figured by Olivier,
and probably it is the type of *bidens* Oliv. The type of *bibens*
Fab. is in Mus. Dom. Tunstall, as stated by Fabricius.
Presumably the name *bidens* indicates the bidentate distal
portion of the outer edge of the front tibiae. The word *bibens*
is, as F. W. Hope pointed out, a printer's error.

Description of Co-type, *Cetonia bidens* Oliv. Form
elongate oval, plano-convex; bronze-green, with brilliant

elytra, hairy, more or less thickly covered with fine yellowish hairs which are short above and long beneath.

The *head* is oblong, a little convex, closely punctured and pubescent; the clypeus is hollowed, its edge is upturned and rounded and emarginate in front.

The *pronotum* is closely punctured, covered with a fine pubescence, and is bluntly triangular, narrowed and almost straight (slightly sinuate) in front, the sides slightly rounded, the front and hind angles rounded, and the base strongly rounded. The *scutellum* is hollowed and triangular, with rounded sides.

The *elytra*, brilliant bronze-green and pubescent, are very short, flattened, marginate, closely punctured, with distinct costal elevations and with the shoulder and apical callus prominent; the sides are vertically deflexed and slightly rugose, the outer margins are gently rounded, broadening out a little about the middle, the apices are rounded, the apical margins slope upwards to the suture, and the apical angles are sharp.

The *propygidium* is exposed and is fringed with long hairs; the large *pygidium* has a minutely imbricate surface and is covered with moderately long and fine hairs. Olivier says, "There is on each side of the posterior end of the abdomen a little oblong whitish mark". The front tibiae have two tooth-like processes on the outer edge at the distal end; the middle and the hind *legs* are long; the hind femora are arcuate, the hind tibiae have two distal spines; the tarsal claws are simple and equal, long and curved.

Length 12 mm.; breadth (across the middle of the elytra) 5 mm. Hab. America (Fab. and Oliv.).

49. *Trichiotinus viridulus* (Fab.)

Coleopterorum Catalogus, pars 75 (S. Schenkling, 1922), Scarabaeidae, Trichiinae, etc., p. 25.

North America.

SYN. *Trichius viridulus* Fab., *Syst. Ent.* Appendix, p. 820, Nos. 5–6 (1775); *Sp. Ins.* 1, p. 49, No. 6 (1781); *Mant. Ins.* 1, p. 26,

No. 8 (1787); *Ent. Syst.* I, 2, p. 122, No. 12 (1792); *Syst. Eleuth.* II, p. 133, No. 12 (1801).
Cetonia viridula Oliv., *Ent.* I, 6, p. 63, pl. 9, fig. 86 (1789).

The two insects in Cabinet A, drawer 1, under label

'*Tr. viridulus*
Fabr. pag. 49, No. 7'

are obviously not that species. It is doubtful whether this type is now in the Hunterian Collection.

50. *Trichius punctatus* Fab.

There are two specimens under label

'*Tr. punctatus*
Fabr. MSS'

in Cabinet A, drawer 1. This species is not mentioned in the published works of Fabricius, and I cannot trace it in the Catalogues.

Super-family STAPHYLINOIDEA

Family SILPHIDAE

51. *Silpha (Necrodes) surinamensis* Fab.

Catalogus Coleopterorum (Gemminger and Harold, 1868), II, p. 720. *Catalogue of the Coleoptera of America North of Mexico* (Charles W. Leng, 1920), p. 87.

North America.

SYN. *Silpha surinamensis* Fab., *Syst. Ent.* p. 72, No. 1 (1775); *Sp. Ins.* I, p. 85, No. 1 (1781); *Mant. Ins.* I, p. 48, No. 1 (1787); *Ent. Syst.* I, 1, p. 248, No. 1 (1792); *Syst. Eleuth.* I, p. 336, No. 1 (1801); Oliv. *Ent.* II, 11, p. 6, pl. 2, fig. 11 (1790).

This species is represented by two co-types under label

'*Sil. surinamensis*
Fabr. pag. 85, No. 1'

in Cabinet A, drawer 6; one is a male and the other is a female, with the wings spread out. They answer the descriptions given by Fabricius and Olivier; and the male corresponds with Olivier's figure of a male in the Banks Collection. I have also compared them with modern examples in the British Museum Collection.

Description of Co-type, *Silpha surinamensis* Fab.

The *male*. Form oblong and flattened, obliquely truncate posteriorly; uniformly dull glossy black, with two reddish yellow wavy bands across the elytra.

The *head* is oblong and punctulate, and marked by a strong constriction immediately behind the very large light brown eyes; the mentum is quadrate; the antennae are eleven-segmented, club-shaped distally, and dull black with the last three segments light brown.

The *pronotum* is sub-orbicular, hollowed in front, and marginate, the lateral parts of the margin strongly developed; the borders are strongly explanate laterally and closely punctulate; the disc is raised and convex and is marked with shallow impressions and faint punctulation. The *scutellum* is large and triangular, with rounded-out sides and the surface punctulate.

The flattened *elytra*, which cover the body entirely, are slightly dilated towards the apices, strongly deflexed at the sides, and marginate; the outer margins are slightly sinuate, the outer angles are a little rounded, the apices are obliquely truncate; on each elytron the shoulder callus and the apical callus are prominent, there are three strongly developed smooth and parallel costae between which the surface is closely punctate, and there is a reddish yellow spot near the middle of the outer margin and a reddish yellow wavy transverse band near the apex. The functional wings are well developed.

The *metasternum* is large and has a small median lobe at the contiguous hind coxae. The *sterna of the six abdominal segments* are rugulose and are fringed with light-coloured hair. The end of the abdomen is pointed. The anterior coxae are large, ridged, and contiguous; the *legs* are stout; the hindmost femora are greatly swollen and have a sharp tooth-like projection on the inner side; the hind tibiae are ridged and are strongly bent in, very arcuate.

Length 25 mm.; breadth (across the elytra apically) 12 mm.
Hab. South America (Fab.), Cayenne and Surinam (Oliv.).

The *female*. The yellow transverse bands on the elytra are shorter, not extending beyond the costae, and the end of the abdomen is broad (truncate).

Family HISTERIDAE

52. *Saprinus detritus* (Fab.)

Coleopterorum Catalogus, pars 24 (H. Bickhardt, 1910), Histeridae, p. 93, *Saprinus detersus* Ill.

SYN *Hister detritus* Fab., *Syst. Ent.* p. 53, No. 10 (1775); *Sp. Ins.*
I, p. 62, No. 10 (1781); (*detricus*) *Mant. Ins.* I, p. 33, No. 13 (1787); *Ent. Syst.* I, 1, p. 76, No. 20 (1792); *Syst. Eleuth.* I, p. 89, No. 28 (1801); Oliv. *Ent.* I, 8, p. 12, pl. 2, fig. 16 (1789).

As stated by Fabricius, the type of this species is in the Banks Collection (British Museum). Olivier described and figured this insect from a specimen in Dr Hunter's Collection. It is noted in the card-index as missing, and I have not been able to trace it.

In *Mant. Ins.* the name appears as "*detricus*", which is probably a misprint. In *Ent. Syst.* Fabricius refers to Rossi, *Faun. Etr.* I, 1790, p. 29, 67. The locality stated by Fabricius and Olivier is New Holland.

Super-family DIVERSICORNIA

Family EROTYLIDAE

53. *Aegithus punctatissimus* (Fab.)

Coleopterorum Catalogus, pars 34 (P. Kuhnt, 1911), Erotylidae, p. 10.

Surinam.

SYN. *Erotylus punctatissimus* Fab., *Syst. Ent.* p. 123, No. 3 (1775);
Sp. Ins. I, p. 157, No. 5 (1781); *Mant. Ins.* I, p. 91, No. 9

(1787); *Ent. Syst.* I, 2, p. 37, No. 10 (1792); *Syst. Eleuth.*
II, p. 5, No. 12 (1801); Oliv. *Ent.* v, 89, p. 469, pl. 2,
fig. 13 (1807).
Coccinella centumpunctatus Herbst, in Füessly.

The specimen under label

> '*Erot. punctatissimus*
> Fabr. pag. 157, No. 5'

in Cabinet A, drawer 4, is apparently the type; it corresponds
with the descriptions given by Fabricius and Olivier, and
with Olivier's figure, and it closely matches modern examples
of the species in the British Museum Collection.

Description of Type, *Erotylus punctatissimus* Fab.
The specimen is defective; the antennae, parts of the legs,
and the abdomen are wanting. Form broadly oval, pointed
posteriorly, and highly convex (gibbous). Dull glossy black,
with reddish yellow and black-spotted elytra.

The *head* is dull black, smooth, short, deeply insunk within
the thorax as far as the prominent eyes; and it forms in con-
junction with the thorax a roughly triangular area.

The *pronotum* is dull glossy black; it is broad, the front is
deeply excavated, the sides are slightly rounded, the front
and hind angles are bluntly prominent, and the base is
angular with a broad lobe in the middle; at each side of the
disc, which is distinctly raised, there are two small foveae,
and between these and each lateral margin there are two
shallow impressions; otherwise the surface of the pronotum
is smooth. The *scutellum* is small, heart-shaped and black.

The *elytra* are highly convex (gibbous), strongly rounded
from the shoulders and pointed at the apices; the surface is
reddish yellow marked with scattered round black and punctured spots, some of which are confluent; the sutural and the
outer margins are narrowly bordered with black; the large
epipleurae are smooth and testaceous.

The *legs* are moderately long, not stout, and are testaceous.

Length 14 mm.; breadth 11 mm.
Hab. America (Fab.), Cayenne, Surinam (Oliv.).

54. *Erotylus histrio* Fab.

Coleopterorum Catalogus, pars 34 (P. Kuhnt, 1911), Erotylidae, p. 23.
Brazil.

Syn. *Erotylus histrio* Fab., *Mant. Ins.* 1, p. 91, No. 3 (1787); *Ent. Syst.* 1, 2, p. 36, No. 4 (1792); *Syst. Eleuth.* 11, p. 4, No. 4 (1801); Oliv. *Ent.* v, 89, p. 468, pl. 2, figs. 12 *a* and *b* (1807).

There are two examples of this species under label

'*Erot. histrio*
Fabr. MSS'

in Cabinet A, drawer 4; they correspond with the descriptions given by Fabricius and Olivier, and Olivier's figure closely resembles them. I have also compared them with modern examples in the British Museum and in the 'Bishop' Collection.

Description of Co-type, *Erotylus histrio* Fab. Form elongate-ovate, highly convex (gibbous). Glossy black, the elytra with irregular reddish yellow bands and blotches and with a scarlet spot on each shoulder and one at each apex.

The *head* is black, short, deeply sunk within the thorax as far as the prominent eyes; the clypeus is hollowed and emarginate; the antennae are pitchy black and have terminal three-segmented clubs.

The *pronotum* is dull glossy black, flattened, moderately broad and marginate except along the middle of the base; the front, which is narrower than the base, is deeply excavated and crescentic, the sides are nearly straight, the front angles are bluntly prominent, the hind angles are sharp, the base is sinuate with a broad middle lobe; the surface is irregular, being marked with shallow impressions, a fovea at each side of the slightly raised disc, and a few punctures at each side of the middle lobe. The *scutellum* is small and heart-shaped, with a median longitudinal ridge.

The *elytra* are highly convex (gibbous) and, being wider at the base than the base of the pronotum, have very prominent and strongly-rounded shoulders; the apices are bluntly pointed, the outer and the sutural borders are marginate; the colour-pattern is glossy black with irregular reddish yellow bands and blotches and with a large scarlet spot on each shoulder and one within each apex, and the surface is marked with scattered black punctures; the large *epipleurae* are smooth and black, with four large reddish yellow blotches.

The underside of the body is glossy black. The *prosternum* is produced behind the front coxae and is united with the episterna, and so the front coxal cavities are closed. The *mesosternum* is short, broad and rectangular, and is marked off from the metasternum by a deeply impressed straight line between the middle coxae. The *metasternum* is marked by a deeply impressed median longitudinal line, and the line of its junction with the basal sternite of the abdomen is a little curved. The five *abdominal sterna* are thinly punctulate, the basal one is a little larger than the others, and each (except the last one) is marked with two oval shallow depressions.

The *legs* are glossy black, moderately long and not stout; the femora are sub-cylindrical with slight ridges; the tibiae are a little expanded distally; the tarsi are four-segmented, the first three segments are flattened, the second is broader than the first and the third is broader than the second, the terminal segment is longer than the others and is cylindrical and it bears two simple curved claws.

Length 23 mm.; breadth 13 mm.
Length 21 mm.; breadth 12 mm. of the smaller co-type.
Hab. Cayenne (Fab. and Oliv.).

55. *Erotylus reticulatus* Fab.

Mant. Ins. I, p. 91, No. 3 (1787); *Ent. Syst.* I, 2, p. 36, No. 2 (1792); *Syst. Eleuth.* II, p. 3, No. 2 (1801).

This type is noted as missing in the card-index of Dr Hunter's Collection, but I have been able to trace it. An un-labelled Erotylid in Cabinet E, drawer 33, corresponds with the descriptions given by Fabricius. Mr Arrow has examined this insect, and he is satisfied as to its identity. We find that this species has been overlooked in the Catalogues; and it will have to retain the generic name of *Erotylus*.

Description of Type, *Erotylus reticulatus* **Fab.** Form oval, very convex; glossy reddish brown, the elytra reddish yellow with an irregular network of chocolate-brown.

The *head* is reddish brown and black, short, convex, sunk within the thorax nearly as far as the prominent eyes; the antennae are pitchy black and have terminal three-segmented clubs.

The *pronotum* is glossy reddish brown with dark brown blotches; it is transverse and is marginate, except along the base; the front is narrower than the base and is deeply excavated, sub-crescentic; the sides are almost straight, the front and hind angles are sharply prominent, the base is sinuate with a broad but slight middle lobe; the surface is a little convex and is irregular, being marked with a central fovea and three foveae triangularly placed at each side and with lighter impressions and a few punctures at each side of the middle lobe. The *scutellum* is wanting in this specimen.

The *elytra* are very convex, but not quite gibbous; a little wider at the base than the base of the pronotum, the shoulders not very prominent and a little rounded, with blunt shoulder angles; the outer sides are rounded in to the narrow blunt-pointed apices; the outer and the sutural borders are marginate; the colour-pattern is glossy reddish yellow with an irregular network of chocolate-brown, and the surface is marked with scattered punctures; the *epipleurae* are smooth and reddish yellow with brownish black blotches.

The underside of the body is glossy brownish black. The *prosternum* is produced beyond the front coxae as a bifurcate process and does not appear to unite with the episterna, so that the front coxal cavities appear to be open. The *meso-* and *meta-sternum* and the *abdominal sterna* are defective, badly fractured, in this specimen.

PLATE 28

Erotylus reticulatus Fab. × 4

The *legs* are glossy reddish brown, moderately long and not stout; the femora are flattened and are slightly ridged; the tibiae are a little expanded distally and are distinctly bent; the tarsi are four-segmented, the first three segments are flattened, the second and third are slightly broader than the first, the terminal segment is longer than the others and is cylindrical, and it bears two simple curved claws.

Length 16 mm.; breadth 10 mm
Hab. Brazil (Fab.).
See Plate 28.

INDEX

The original names are printed in **bold** type.